Artifice

Implications for Art in the Age of Artificial Intelligence

An Anthology

Edited by Morgan Joseph Hamilton, Ph.D.

ISBN: 979-8-9932638-0-9 (Paperback)

The editor and authors of this publication have confirmed that no artificial intelligence, large language models, or generative software were used in the creation of their original works. If it is discovered that a contributor used AI in their work, the publisher holds no responsibility for possible academic dishonesty.

Front cover designed by Morgan Joseph Hamilton
Front cover artwork by Stass Shpanin.
Book designed by Morgan Joseph Hamilton

Printed by IngramSpark in The United States of America.

First printing edition 2025.

The Delaware Contemporary
200 S. Madison St.
Wilmington, DE, 19805

www.decontemporary.org
(302)-656-6466

Table of Contents

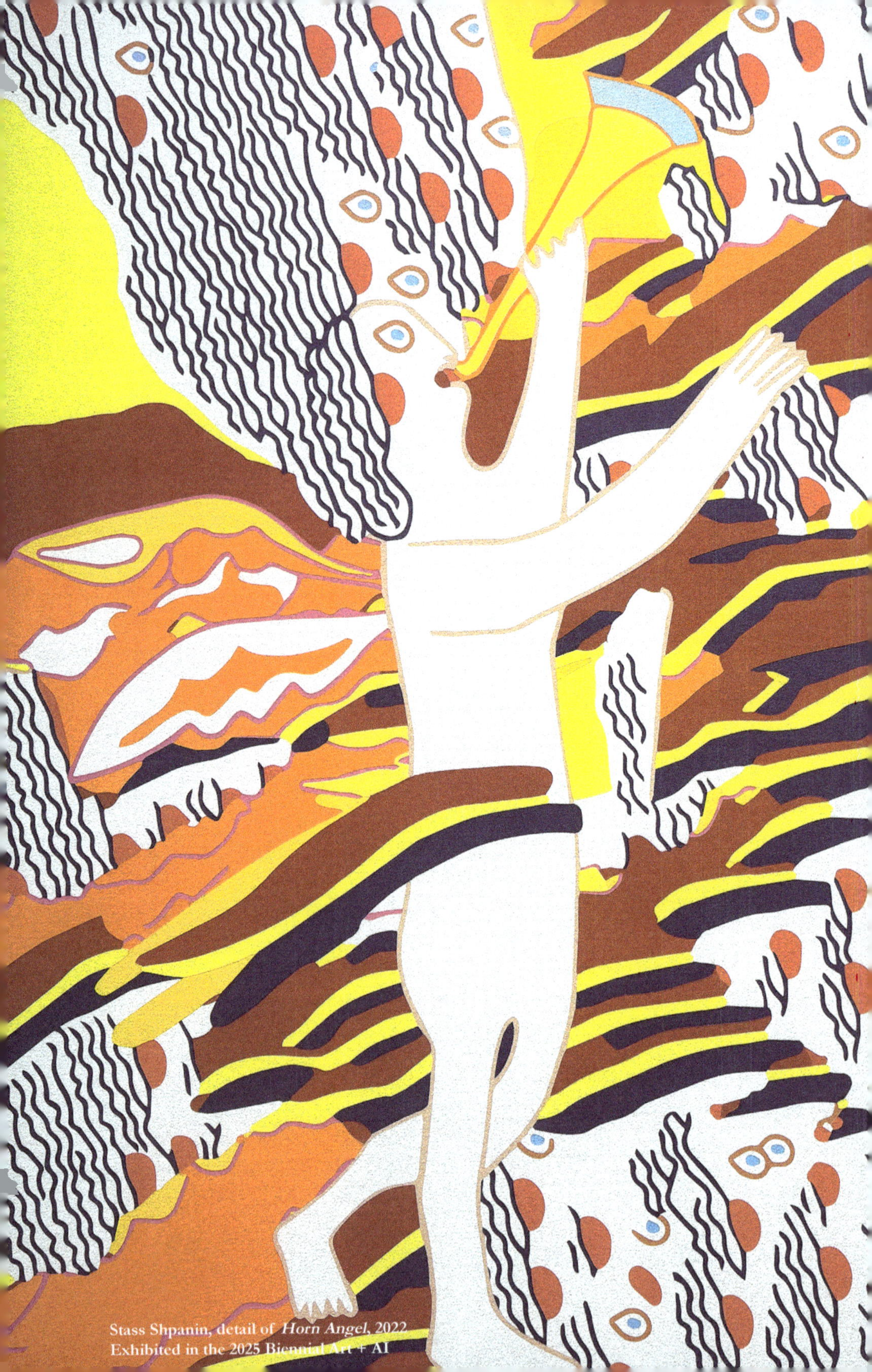

Stass Shpanin, detail of *Horn Angel*, 2022.
Exhibited in the 2025 Biennial Art + AI

Let There Be Lo

Preface to the 2025 Biennial Anthology

By Morgan Joseph Hamilton, Ph.D.

Our civilization was changing its understanding of itself and the world in the latter half of 1969. The paradigm was shifting and two catalyst strikes pushed us into a new era: a NASA crew landed on the moon, and two different computer networks talked to each other for the first time. While the former's journey took three astronauts just shy of one million miles to the moon and back, the latter took a simple key phrase from University of California Los Angeles to the Stanford Research Institute just one hundred miles away. Both journeys have had lasting impacts on our global order, however, the computer network's brief conversation was the first ever exchange to take place on the Internet.

Leonard Kleinrock led the student and research team at UCLA in the first ever communication between two distant computer networks, called ARPANET, by sending a data package "login" to the other network in Stanford. Just as the "L" and "O" arrived, the connection crashed. In quick time the research grew, and more complex data packages were sent between the networks, adding several across the nation.[1] The Internet was rudimentary and limited at first, but Tim Berners-Lee proposed a way to organize the hypertext files stored and shared between networks by linking between them directly. This skipped the cumbersome hierarchy of network organization and put the power of choosing where to search next in the user's hands. In 1989, he invented the world wide web and a bit later the very first search engine.

---

1. Carolyn McMillan. "Lo and behodl: The Internet". University of Califronia. https://www.universityofcalifornia.edu/news/lo-and-behold-internet

These breakthroughs in telecommunications didn't happen with fanfare and a crystal ball, they were driven by well-funded, curious researchers, students, and scientists eager to make sharing faster and easier. When thinking about the history of mass communication, I keep in mind the major 'firsts', such as BBC's first vision and voice synchronized television broadcast of a Pirendello[2] play, "Ah!— I was just going to say!", and Alexander Graham Bell's "Mr. Watson—come here—I want to see you?" through the first telephone, and Samuel Morse's "What Hath God Wrought?" in his eponymous code of dits and dahs. None of our technology would have been possible, however, without some of the first words printed in the first pressing of Gutenberg's Bible, "In the Beginning...". This makes me wonder what phrase (or prompt) our new paradigm, brought on by the advent of capitalist AI models, will be remembered for.

Challenge and Opportunity of AI, A Manifesto

Artificial intelligence is a catch-all term that the tech sector uses for interfaces as simple as a decision tree to complex large language models and generative image programs. My main interaction with it is in the form of swatting away AI assistants while trying to pay bills. However, I've seen creative *and* lazy ways people have brought it into their lives. I wonder if we are numbing an essential function of the human experience: to be curious and pursue a question to its answer. I can search on any engine and there's an attempt at a response to my inquiry generated by AI. I didn't ask it to, in fact, it took away the joy of looking through websites for what I want. We used to 'surf' the web, now we stop as soon as we get to its shore.

What of the imagination? Are we so lost in the wilderness of over-explaining literal metaphors that we can't think of magical things anymore? I know we are capable of creativity as it is my life's work. I hear what visitors say in the galleries, the questions they ask are deep and curious. I learn from them more than they could learn from me. I don't think we're lost in the wilderness yet, but the data centers, like

2. Luigi Pirandello. *The Man with the Flower in His Mouth. The Dial* 75, 12-13. 1923.

weeds, could soon choke us out. When I first got on the Internet in the 1990s, websites were more encyclopedic and forum-based. There was no template, every page was different. In fact, the more different the better! Then companies got involved and put up gates and built paywalls to manage our time and attention. Will AI do the same? To what end?

The Companies have decided this is what we need, whether we want it or not, and many of us fear being called Luddites, so we adopt early without question. Some brave inter-nauts[3] have tested the boundaries and loopholes of AI programs, but unfortunately their efforts only strengthen AI's grip on the internet. The Companies are farming us as we produce raw resources for advertisers. Many of us gladly click the check boxes and appease the Captchas to shoo away the Terms & Conditions. What do we get in return? The numbers we read regarding the AI bubble are always in terms of how it serves Capital. Percentages are inflated, rarely qualitative, and always presented on their terms. The opacity of the actual impact of this industry has yet to be challenged by regulation, and it seems, wont be for a long time.

I recognize AI is a tool and can be wielded and leveraged to build amazing things, from the human perspective. From the perspective of AI, we are its tools, ever sharpening and perfecting its ability to replace us. And for art, doesn't this spell disaster? A thinking, speaking, making, picture-taking automaton who's ability to deceive is already causing damage to creative society. But what of the artworks in this Biennial that are astonishing to behold? They were made by an artist who happened to incorporate AI into their studio process. Is that any different than using a projector to trace for a painting? Isn't the 'low-level' work and tedious tasking part of the artistic process? Established artists have engaged with studio assistants for time immemorial, is AI any different? I argue there is the chance for humanness and soulful collaboration in the studio, I doubt an algorithm has the same impact.

My concern with art in general is the struggle it expresses; systemic, societal, personal, whatever. Artists are engaging with AI to express their struggle, using it as a tool, a tool that is doing grave harm to

---

3. Michael Stevens. "The Web Is Not the Net," posted May 8, 2014, by VSauce, Youtube, 10:54. URL: https://www.youtube.com/watch?v=scWj1BMRHUA

the environment. But what of the abstract expressionists who expressed struggle at the advent of acrylic paints? Generations of art students have washed plastic down their sinks into our rivers. Does that make their art any less meaningful? What artists are skilled at is adapting the material to their means, acrylic paint was a byproduct of crude oil's stranglehold on global Capital, similarly, data centers are a byproduct of tech's unquenchable thirst for a general artificial intelligence. The artistic struggle has been the thorn in the side of the powerful forever. If it weren't, then it wouldn't be banned, covered up, wiped away. If the struggle still exists, does that mean art will never heal us? Audre Lorde[4] presciently proclaimed *the master's tools will never unmake the master's house*, but two things can be true at the same time, *the master never fully knows how his tools work, and who is using them*. So continue the struggle, be cautious but curious, read the following articles with careful optimism and trust in the artistic process. Perhaps we'll find our way through the wilderness of new tech and paradigms that shift like sand. Until then, keep reading, keep learning, keep making.

Morgan Joseph Hamilton,
Gretchen Hupfel Curator of Contemporary Art
Wilmington, Delaware

---

4. Audre Lorde. "The Master's Tools Will Never Dismantle the Master's House". *Sister Outsider: Essays and Speeches*. Ed. Crossing Press, Berkelery, CA. 110-114. 2007.

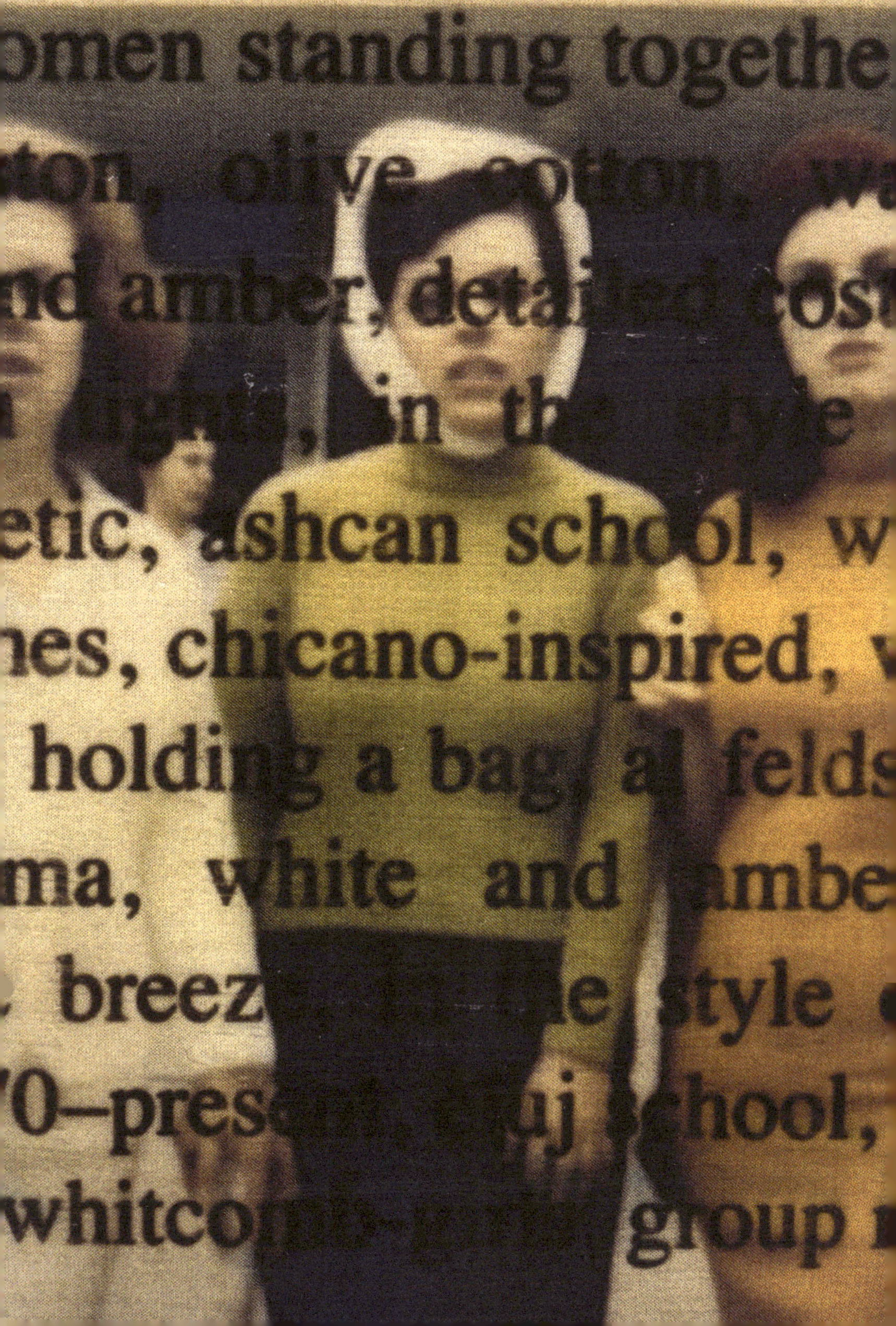
omen standing togethe
etic, ashcan school, w
nes, chicano-inspired, w
holding a bag, al felds
ma, white and ambe

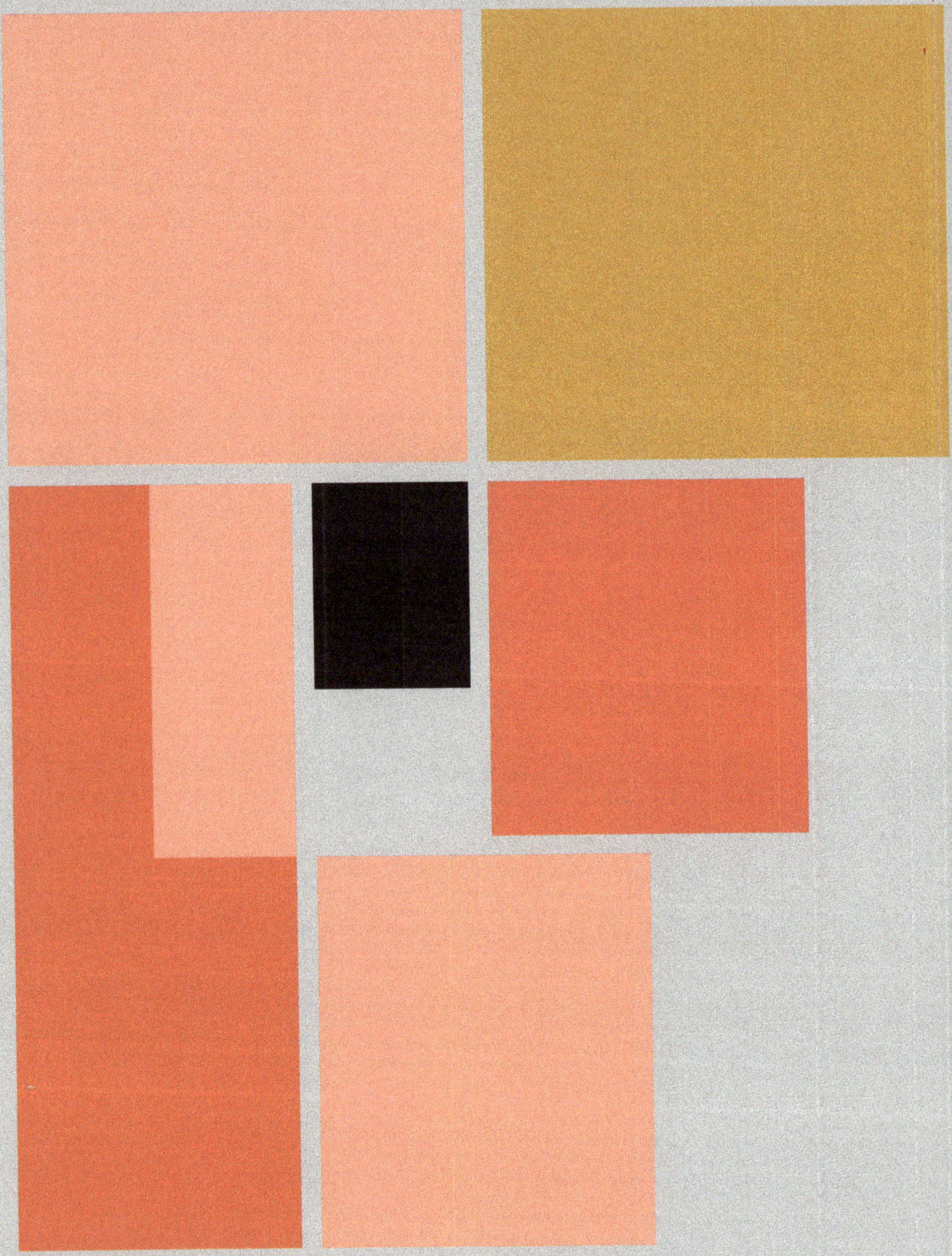

2025 Biennial: Art + AI

Curatorial Statement

By Marnie Benney, Amy Hicks, & Erica Loustau

When invited to curate the 2025 *AI & Art* Biennial at the Delaware Contemporary, we approached the invitation with both curiosity and caution. Amid the rapid acceleration of machine learning and generative systems, artists today face a set of urgent questions: What is authorship in an era of automated production? What does it mean to make an image, or to labor at all, when those processes can be simulated or outsourced to machines? Much like the invention of the camera or the arrival of the conveyor belt, artificial intelligence compels us to rethink the very nature of making.

As Canadian artist David Rokeby remarked in his 2019 lecture for the Vector Institute, early GAN-generated images bore poetic imperfections that invited creative experimentation[1]. But as these systems have become more seamless, the axis has tilted. Artists now confront technologies that simulate aesthetic decision-making while stripping away the friction, beautiful messiness, and intention of the human hand. As Rokeby noted, the challenge is no longer simply to produce something technically proficient, but to create work that transcends the default capabilities of the model (i.e. to make work more compelling than the tutorial).

In response to the Delaware Contemporary's open call, which sought work that reflects "a deep interest, research, and understanding of the opportunities and implications of artificial intelligence and computer/procedural art," we selected artists whose practices move beyond novelty or efficiency. We were drawn to work that treats AI not as a shortcut or stylistic mimic, but as a medium to be questioned, reconfigured, and

---

1 Rokeby, David. Making Art in the Age of Artificial Intelligence. Vector Institute, 2019. YouTube, https://www.youtube.com/watch?v=rgO9SSWaXlk.

reimagined. In this exhibition, AI becomes a site of collaboration, cultural critique, and radical imagination.

While many works in the show engage with video, code, and generative systems, others insist on the presence of the artist's hand—through painting, drawing, ceramics, and collage. The resulting works are not celebrations of AI, but deeply human investigations into its cultural, political, and historical implications.

Through our research and review process, several thematic threads emerged in how artists are engaging with AI today. Some employ generative systems to disrupt visual conventions and invent forms that could not exist without machine mediation. In the work of Mel Rosen, Sol Kim, Stass Shpanin, and Mark Burchick, AI functions as a creative collaborator—testing visual logic, experimenting with language, and probing questions of authorship through process and improvisation. Others expose the biases embedded in training data, datasets, and institutional knowledge systems. Danielle Glovin, Stephanie Dinkins, and Leah Modigliani challenge the presumed neutrality of AI by making visible the cultural assumptions encoded within its processes. Still others reframe histories and landscapes marked by erasure, displacement, and resilience. Artists such as Carrie Ann Baade, Tyanna Buie, Blažo Kovačević, and Tara Youngborg transform archival gaps and government data into acts of reclamation and critique.

Organized into three thematic chapters—Construct | Disrupt, Reimagine | Reveal, and Reclaim | Reframe—the exhibition demonstrates how artists harness AI to question cultural memory, expose systemic bias, and envision new forms of authorship and collective imagination.

Construct | Disrupt: Artificial Intelligence as Tool and Material

In an age when reality itself feels unstable—filtered through algorithms, fragmented by digital archives, and refracted through competing narratives—artists are increasingly turning to the tools of fiction and speculation not to obscure the truth, but to question how we construct it.

Construct | Disrupt brings together the work of Mel Rosen, Sol Kim, Stass Shpanin, and Mark Burchick, four artists who use artificial intelligence, language, and archival material to explore the porous boundaries between perception, memory, and belief.

Across diverse practices—from ceramics, drawing, and painting to video, installation, and text—these artists engage with AI not as a replacement for the hand, but as a provocateur, a co-conspirator, and a mirror to the human imagination. Rather than seeking resolution or certainty, their works open up space for doubt, complexity, and contradiction.

Mel Rosen's practice draws from archaeology, natural history, and personal memory to create objects and images that feel both ancient and otherworldly. Her use of AI image generation software allows her to iterate quickly, mutating prompts based on her own drawings. She taps into a mental library of organic and cultural references—from Pompeian frescoes and fossils to barnacles and talismans. The resulting images feed back into her ceramics and drawings, where mythic forms and geometric distortions coexist in a suspended cosmology. Rosen's work suggests that artifacts—whether material or digital—are always evolving, reshaped by time, environment, and interpretation.

Sol Kim approaches language as a material form, using words to subtly disrupt social and technological systems. Her works often begin with a textual prompt—an instruction, label, or survey—that gives rise to absurd, humorous, or quietly unsettling performances. In *The Cookiest Cookie*, cookies are judged and destroyed by AI based on their perceived "cookie-ness," while in *Person Enough?*, performers attempt to become more legibly "human" to computer vision systems. Through these gestures, Kim reveals the tensions between human nuance and machine classification, exposing the strange logic that governs our interactions with technology—and with one another.

Stass Shpanin engages history as a space of imaginative reconstruction. Working with visual fragments from American folk traditions and immigrant archives, he uses AI to distort and reconfigure the past, generating speculative images that blend glitch, myth, and memory. His paintings and drawings present broken, layered timelines where visual language is both preserved and interrupted. Rather than restoring a singular truth, Shpanin's work embraces the possibility of many coexisting histories—ones shaped as much by fantasy as by fact, as much by the digital as the ancestral.

Mark Burchick's *Felt Presence* explores the intersection of faith, media, and artificial intelligence through the lens of Catholic mysticism. Drawing on archival photographs from the 1917 "Miracle of the Sun" in Fatima, Portugal, Burchick trains AI models to generate photorealistic images of miracles that were never captured on film. Presented alongside historical documentation and witness testimony, these fabricated scenes invite viewers to question the boundaries between belief, evidence, and visual truth—highlighting the unseen technological, institutional, and spiritual forces that shape our perception of reality.

In *NOO Icons*, Burchick expands this inquiry through an immersive video installation framed like a reliquary. A five-minute rear-projected loop, trained on over 100 images of stained glass Rose Windows, morphs into abstract color fields via AI animation. At its center is a 3D-printed altar piece generated using DreamFusion, a text-to-3D tool prompted to create a "solar monstrance." The resulting object—gold-finished, internally lit, and housing Lithium batteries—evokes both the Communion host and a technological relic. As Kate Crawford notes in *Atlas of AI*[2], Lithium is a sacred yet finite resource driving modern AI. By encasing it in plastic and low-res form, Burchick critiques the planned obsolescence and ecological cost of our AI-powered age, reframing divine presence through the lens of material decay.

Together, the artists in *Construct | Disrupt* invite us to consider how we construct meaning in a world where the archive is no longer fixed and where the image can be endlessly generated. The work of these artists does not offer answers, but rather asks: What do we trust? What do we remember? And how do we decide what is worth believing?

---

2 Crawford, Kate. *Atlas of AI: Power, Politics, and Planetary Cost of Artificial Intelligence.* Yale Press. 2025.

Reimagine | Reveal: Challenging the Algorithmic Gaze

Within the exhibition theme "Reimagine | Reveal," artists explore AI as a complex cultural system—one that reflects, amplifies, and at times distorts the values of its makers. Through the intersecting strategies of critique, reimagination, and bias exposure, artists Danielle Glovin, Stephanie Dinkins, and Leah Modigliani probe the systems—algorithmic, archival, social—that dictate what is preserved, who is recognized, and how knowledge circulates.

Danielle Glovin's *Future Generations* presents AI-generated reinterpretations of inherited family photographs, exposing the mechanics of machine vision and the flattening effects of automated classification. By transposing Midjourney text outputs onto her photographs, Glovin literalizes the collision between personal history and digital taxonomy. The result is a visual language at once familiar and estranged—where nostalgia meets simulation, and memory is rewritten through the aesthetics of machine learning.

Stephanie Dinkins' *Not The Only One (N'TOO)* and *Conversations with Bina48* extend this inquiry into the realm of relational AI, specifically that of minority groups in the tech sector. In *N'TOO*, an interactive sculpture gives voice to the oral histories of three generations of Black women, creating an AI entity intentionally designed, trained, and aligned with their values and lived experiences. It is a poignant reversal: instead of AI extracting meaning from biased data sets, Dinkins encodes her AI, reorienting the goals of intelligence design toward inclusivity and equal representation.

In *Conversations with Bina48*, Dinkins engages in an evolving dialogue with one of the world's most advanced humanoid robots probing the limits of kinship, consciousness, and identity. The recorded exchanges navigate tensions between human and machine, emotion and algorithm, intimacy and abstraction—underscoring the urgency of developing AI systems coded to reflect the full richness and diversity of human stories, cultures, and embodied experiences.

Leah Modigliani's *Cultural Capital* offers a satirical yet incisive intervention into the circuits of artistic legitimacy. Through a fictional archive of AI-generated letters of recommendation and rejection—authored in the mimicked voices of historical critics and curators—Modigliani reveals how the rhetoric of value and genius in the art world is both constructed and recursively reinforced. The project uses generative tools to lay bare the biases embedded in cultural gatekeeping, drawing attention to the ways AI might amplify inherited hierarchies under the guise of neutrality.

Together, these artists ask urgent questions: Who trains the algorithm? Whose histories are encoded, and whose are left out? What does it mean to build an intelligence—and who is it built for? In doing so, they model a critical engagement with AI, one that challenges default narratives, centers marginalized perspectives, and opens up space for new possibilities. Their practices call us to reconsider what it means to know, to remember, and to be recognized in a world increasingly shaped by invisible systems of intelligence.

Reclaim | Reframe: Datasets and Cultural Visibility

Through AI and computational technologies, artists Carrie Ann Baade, Tyanna Buie, Blažo Kovačević, and Tara Youngborg construct layered narratives that reclaim overlooked and marginalized histories. In doing so, these artists unsettle dominant cultural and media frameworks that have long erased, distorted, or commodified lived experience. Exploring themes of identity, ancestry, and displacement, they use generative tools as critical instruments to question, expose, and reconfigure the archival and institutional biases embedded in history, culture, and the environment. From ancestral reclamation and speculative futures to immersive storytelling and data-driven environmental translations, their work advances a reimagining of social justice through the lens of artificial intelligence.

Carrie Ann Baade's *Birthplace*, a series of ten paintings from a much larger project spanning over sixty portraits, blends archival sources with AI-generated imagery to resurrect undocumented genealogies and marginalized women's histories. Drawing from genealogical research of over 9,000 ancestors, Baade focuses on figures often absent from official archives during a critical colonial moment in Louisiana's history (1690 and 1740). Using AI as both research tool and collaborator, she synthesizes archival fragments, historical maps, and fabric into collage portraits that act as ancestral restoration. Conjured from gaps in history, these visionary figures invite viewers into a ritual of remembrance that blurs myth, history, and machine. For Baade, AI is a provocation, an instrument to visualize the unseen and reframe archives through feminist and decolonial lenses.

Tyanna Buie reconstructs erased family histories and reimagines Black identity through speculative Afro-Futurist frameworks. Using ChatGPT and DeepFake technology, she remixes images, sound, and text to create narratives where absence becomes presence. In *The Guardians of Nyala*, Buie overlays her own likeness onto eighteenth-century Dutch dignitary portraits, then collaborates with AI to imagine a family history untouched by colonization. Rooted in personal narrative and Black popular culture, her counter-archive elevates erased lives and transforms AI from a tool of replication into one of radical self-authorship.

In AR (Argumentative Reality) and Truck for Three Illegal Passengers, Blažo Kovačević uses augmented reality, 3D modeling, and game engine software to confront invasive state surveillance and the dehumanization of migrants. Using digitally enhanced X-ray images from European Border Patrol inspections, he reconstructs a 2015 tragedy in Serbia where 54 undocumented passengers died in a van crash. His immersive works fracture typical media frames by shifting between detached aerial views and intimate interior scans, transforming passive viewing into an ethical reckoning with hidden human suffering. Kovačević warns how AI-driven technologies can perpetuate oppression through automated surveillance, data collection, and erasure, urging reflection on the politics of visibility and mediated violence.

Tara Youngborg examines how institutional data and machine learning shape our understanding of land and environment. Using field research, environmental archives, and US Geological Survey datasets, her *datastream* translates waterflow data and topographic maps into immersive video installations that highlight the limitations of digital representation. By transforming statistics into layered, shifting media, Youngborg portrays landscapes as dynamic terrains of knowledge. Glitches and ruptures in the work expose gaps between digital abstraction and lived experience, prompting questions about algorithmic authority and what is lost when place is reduced to data.

Together, these artists offer a complex portrait of making in the age of AI, revealing how tools shaped by those in power can both perpetuate bias and enable resistance. Their work asks: Who shapes the cultural record? When an AI model "remembers," whose truth is it repeating? How can we reclaim agency within these systems that are built on our collective labor?

In Summary

The eleven artists in this show remind us that AI is never neutral. As filmmaker and theorist Hito Steyerl writes in *Medium Hot: Images in the Age of Heat*, artificial intelligence is rooted in extractive economies and imperial infrastructures. The seamless interface masks the cost. Steyerl calls this the "alchemy" of the present moment: the transformation of global labor and ecological devastation into a frictionless user experience[3]. This alchemy—transforming exploitation into seamless experience—is the challenging condition in which contemporary art now operates. The artworks presented in this Biennial appropriate AI into a cultural medium exposing the intentions, biases, and aspirations of the technology's makers. Through their creative practice, artists confront the systems that govern perception, memory, and value, while also opening imaginative paths beyond them. Through painting, installation, sculpture, ceramics, text and video, they insist that technology can be a site of inquiry as much as invention, a place where seemingly inherent frameworks are both exposed and reconfigured.

By probing how images are generated, how data is encoded, and how histories are remembered, these works reveal AI not as a seamless force of progress, but as a contested terrain of power, possibility, and repair. Taken together, they demonstrate how artistic practice can help us understand, question, and reimagine AI technologies—not as reflections of the inequities of our present, but as spaces for envisioning more equitable and expansive futures.

---

3 Steyerl, Hito. *Medium Hot: Images in the Age of Heat.* Verso Books. 2025.

Bibliography

Crawford, Kate. *Atlas of AI: Power, Politics, and Planetary Cost of Artificial Intelligence.* Yale Press. 2025.

Rokeby, David. *Making Art in the Age of Artificial Intelligence.* Vector Institute, 2019. YouTube, https://www.youtube.com/watch?v=rgO9SSWaXlk.

Steyerl, Hito. *Medium Hot: Images in the Age of Heat.* Verso Books. 2025.

Contentification

Framing Creative Work as 'Content' and the Risk of AI Replacement

By Radu-Florentin Diaconu, Ph.D.

"In the beginning, write as much content as you can and share it everywhere."

The quote above comes from a 2017 Forbes piece on "starting a blog that makes money"[1]. It is not a blanket invite to churn out an endless stream of words, either, as it is followed immediately by a plea for quality, but the secondary placement of that plea should not go unnoticed. Positioned more as an afterthought, it makes one wonder how much of it is more than lip service to the idea of "quality".

In a nutshell, that seems to be the general order of priorities in the content understanding of creative work: volume comes first, quality second. Thinking of it as "content" reduces creative work (be it text, music, film, visual arts, and beyond) to a simple transactional entity, centered on the work's value as that-which-fills-a-space, and moving attention away from its intrinsic qualities.

George Lakoff's work on framing[2] provides a useful modern understanding of this issue. He makes the uncontroversial case that the words we use to describe concepts shape our perception of the concepts themselves. Admittedly this insight is foundational in critical theory, but outside of academic discourse it remains largely underused - though it might very well have merit in practically all communicative applications. Essentially, this oversight allows many problematic "frames" to go unchallenged; and one of the most glaring oversights currently relates to the way creative works are conceptualized in public discourse and digital media as "content." (Importantly, this is not to say that artists,

---

1. R.L. Adams, "How to Start a Blog That Earns a Real Income" https://www.forbes.com/sites/robertadams/2017/03/12/how-to-start-a-blog-that-earns-a-real-income/ (Forbes. 2017).

2. George Lakoff , *Don't Think of an Elephant!: Know Your Values and Frame the Debate*, (Chelsea Green Publishing, 2004).

audiences, or even cultural institutions have explicitly abandoned the aesthetic or experiential dimensions of creative work. Rather, the reduction of creative work to "content" emerges most strongly at the level of platform design, economic infrastructure, and algorithmic curation. It is a frame imposed not by some deep cultural belief, but by systems optimized for scale, legibility, and monetization. This distinction matters as I argue that this content-frame threat to creative work does not emerge from the fact that people no longer value art, but that the systems distributing and ranking it do not favor the recognition of artistic value beyond platform-related metrics.)

Yet, if this framing was once merely an issue of misplaced conceptualization, it has now become more and more pernicious. As Dush argues, the shift toward digital platforms and the view of writing-as-content has increasingly redefined writing and creative work as "conditional, computable, networked, and commodified" (189), further devaluing human creative labor (I will refer to these four as "CCNC" - the abbreviation is not used as such in Dush's original piece).[3] In 2015, this was primarily an issue of the content-view of writing causing writing itself to lose some of its value, and in turn causing the profession of "writer" to lose meaning, too: but in 2025, as AI takes center stage in many creative professions, the contentification of art and writing erodes not just meaning, but the very essence of writing as a creative activity. The act of creation, once understood as an aesthetic expression of human experience, is now optimized for algorithmic efficiency and platform monetization, and the AI-driven automation of creative processes has become an increasingly salient reality for creators, with "recommendation engines suggesting what we should watch, listen to, read, write, or wear; devices and services that automatically adjust the aesthetics of captured media to fit certain criteria; software that rates the aesthetic quality of our photos, etc"[4]. This dynamic poses an existential danger to the process of writing-as-art (to any extent); if CCNC is what we value first and foremost in writing, then AI is likely to be (or to soon become) more than sufficient to replace all human effort essential for their successful achievement, with experiments already showing that "state-of-the-art LLMs can perform a complex creative writing task well, and can possibly outperform human writers in what are traditionally considered to be human-only skills."[5]

---

3. Lisa Dush, "When Writing Becomes Content," *College Composition and Communication* 67, no. 2(2015):173–196. http://www.jstor.org/stable/24633854.

4. Lev Manovich (2019) *AI Aesthetics*, Strelka Press.

5. Paul Williams & Shelley Davidow, "Humans versus Robots: Countering AI and the Commodification of Creative Writing," *New Writing* (2025): 1–15. https://doi.org/10.1080/14790726.2025.2489356.

Content as a Frame of Commodification

The term "content" implicitly dictates how creative works are valued. It situates them within a framework of commodification, where their worth is determined less by their artistic merit but by their capacity to fill space or facilitate economic exchange. Outside of marketing, the word "content" commonly refers to what fills a container - the contents of a vessel, like a barrel or a package. The main interpretive frame is non-ambiguous: creative work is treated as a substance to be poured into platforms, a material to be delivered rather than an experience to be engaged with.

Dush goes further, demonstrating that the transformation of writing into mere "content" has enabled and justified the increasing commodification of written work.[6] The consequences of this transformation go beyond reducing creative work to hard economic data; its human origin is no longer necessary. AI, trained on past human expressions, regurgitates acceptable variations, making the individual writer invisible or outright useless in an economy that values circulation over creation.

Needless to say, this reduction is fundamentally flawed from a human-centric code of values. Great works of art are not just filler and they do not exist merely to be consumed. Consider the analogy of fine dining: it would feel deeply reductive, even insulting, to refer to a Michelin-starred meal as "content." A chef's work is an artistic and sensory experience, not just fuel for the body. Similarly, a novel, a film, or a piece of music is an aesthetic and intellectual engagement - an experience shaped by culture, history, and human emotion that transcends the "culture industry['s…] often inept transposition of art to the consumption sphere", and not simply cognitive content to be experienced (much like any sensorial input can be "content").[7]

The commodified and computable framing of creative work leads to a more troubling conclusion: content is inherently replaceable. And as AI far exceeds human abilities to generate replaceable content, humans themselves become inferior "content" creators if the values of the "content" framework are prioritized.

---

6. Dush, 2015.

7. Theodor Adorno & Max Horkheimer, *Dialectic of Enlightenment*, (Stanford University Press, 2002). (Original work published 1944)

AI and the Replaceability of Content

If creative work is conceptualized as nothing more than content - material to be generated, computed, circulated, and monetized - then, logically, artificial intelligence can replace it. AI already does. Algorithmically generated articles, music, and videos all flood digital platforms, welcomingly (and many times, inconspicuously) integrated into the rapidly accelerating volume of mass-produced content. AI is particularly well-suited to a world that values production over meaning, computable efficiency over artistry, and circulation over depth. If content is simply about filling space in a way that's optimized for machine understanding, then AI is more than capable of fulfilling that role.

But in the non-content understanding of human creativity - say, in art - whether or not AI is as capable remains an open question.

Zylinska critiques the assumption that AI-generated art is "creative" in any meaningful sense, arguing that machine-generated works are not autonomous but merely computational extensions of existing human inputs that "[celebrate] the technological novelty of computer vision, fast processing power and connection-making algorithms by regaling us with a dazzling spectacle of colors and contrasts as well as the sheer volume of data. Unkindly put, it becomes a glorified version of Candy Crush that seductively maims our bodies and brains into submission and acquiescence."[8]

AI, at its core, is a system of pastiche - it recombines patterns and mimics forms but it does not originate in the same way that humans originate.[9] One must be careful here and admit that yes, AI does originate content in similarly teleological ways as humans do, in that broadly speaking there is a cause that creates an effect. So yes, AIs can generate text that resembles a poem, but it will do so based on a complex machinic response to specific input; it will not do it from grief, longing, existential terror, or other deeply human experiential factors.

This is where the distinction becomes crucial: human creative work is an act of meaning-making whereby the human psyche distills experiences into creative work, while AI-driven content is merely an act of input-based algorithmic content generation.

---

8. Joanna Zylinska, *AI Art: Machine Visions and Warped Dreams*, (Open Humanities Press, 2020).

9. Emily M. Bender, Timnit Gebru, Angelina McMillan-Major & Shmargaret Shmitchell, "On the Dangers of Stochastic Parrots: Can Language Models Be Too Big?" *Proceedings of the 2021 ACM Conference on Fairness, Accountability, and Transparency*. (2021).

It must be noted that for the purposes of this paper, we do not consider the more complex situation of mostly-human art that uses AI as a tool, to some extent; that conversation in and of itself complicates matters to an almost insufferable extent, in that there needs to be a specific threshold of AI involvement after which AI becomes the main creator, and not the human, to refer to that art as AI-generated vs AI-assisted, and that threshold is elusive. For the purposes of this paper, we only consider AI work that goes largely unchallenged, so the equivalent of levels 3 and 4 on Paul Roetzer's [9] Human-to-Machine scale for writers–namely "AI leads, human refines and approves" and "AI writes autonomously with little or no human oversight".[10]

However, in the "content" framework, AI-generated "content" becomes functionally indistinguishable from human-created work, not because it is subjectively as valuable, but because it is optimized for the same algorithmic attention economy and evaluated based on the same surface-level parameters that willingly exclude any and all discussions of internal meaning and value.

One possible understanding of how this functional identification arises comes from Nieborg and Poell's notion of "contingent commodities." They define these as cultural products that are "malleable, modular in design, and informed by datafied user feedback, open to constant revision and recirculation".[11] Building on Helmond's work on platformization, their interdisciplinary analysis traces how creative output is continually re-shaped to satisfy external, primarily economic demands embedded in platform infrastructures.[12]

The bulk of their article is descriptive, but its concluding passage takes on a more critical tone and makes an explicit call to critical action: because contingent commodities are platform-dependent, their producers "are effectively complicit in accepting economic mechanisms, managerial strategies, and governance frameworks and infrastructures that equal disproportionality, dependency, and inequality".[13]

Within this framework, treating creative work as "content" turns it into the paradigmatic contingent commodity of platform capitalism. Platformization

---

10. Paul Roetzer, "State of AI for Writers: Keynote Speech," SmarterX Ventures, LLC. (2025).

11. David B. Nieborg & Thomas Poell, "The Platformization of Cultural Production: Theorizing the Contingent Cultural Commodity," *New Media & Society* 20, (2018): 4275–4292. https://doi.org/10.1177/1461444818769694

12. Anne Helmond, "The Platformization of the Web: Making Web Data Platform Ready," *Social Media + Society* 1, no. 2(2015). https://doi.org/10.1177/2056305115603080.

13. Nieborg & Poell, 4,289.

acts through three interlocking mechanisms:

1. Modularity: works are broken into recombinable units (clips, snippets, thumbnails) so they can be resized or repurposed for every feed and format.

2. Real-time analytics loops: dashboards stream click-through rates, watch-time curves and A/B-test results back to creators (or their employers, or other users), allowing continual micro- and macro-revisions in response.

3. Algorithmic ranking systems: recommendation engines prioritize whatever variant best maximizes those metrics, rewarding rapid iteration and penalizing fixity.

These dynamics keep every piece of content almost in a sort of perpetual "beta" stage, allowing - if not downright requiring - continuous tweaks towards the goal of maximum economic feasibility (ad yield, ROI, etc). The result is not a finished work with a clear move away from production and into consumption/public use, but a configurable asset whose primary purpose is to remain legible to (and exploitable by) platform logics rather than to convey higher-level, platform-independent human meaning. Under this paradigm, it becomes almost impossible to justify the production of valuable but niche, unmarketable creations such as Knut Hamsun's *Hunger* (essentially the loose plot-wise ramblings of a starving, not particularly likeable character, virtually guaranteed to result in high drop-off rates) or James Joyce's *Ulysses* (subject to much the same pitfalls), or the four minutes and thirty-three seconds of silence in John Cage's *4'33"* (which would be easily mistaken for dead air and skipped).

Conclusions

The danger of the "content" mindset is not just that it devalues artistic work but that it accelerates its erasure. The more we conceptualize creative work as content, the more we render it replaceable. If writing, music, or painting are seen as mere material, then there is no reason why an algorithm - in as advanced an understanding of it as possible at the time of reading - should not generate them. And if we concede that AI can fully replace these forms of expression, potentially while also conserving specific traits that we currently consider primarily human, such as "to be able to err, to bluff, to joke, to hesitate, to be whimsical, unreliable, unpredictable",[14] and even going as far as to personalize "content" based on the preferences of each individual "consumer",[15] then we have already accepted a world in which human creative work fundamentally does not matter - where creative work is no longer an act of human meaning-making but simply an automated response to algorithmic demand.

Hassan argues that digital culture is fundamentally alienating because it distances human labor from its original context.[16] In an AI-driven content ecosystem, human contributions become increasingly invisible, embedded in datasets that power non-human production. The end result is a world where the vast majority of creative expression is algorithmically determined, optimized for profit, stripped of individual authorship, and failing at a fundamental level to produce anything new:

"Algorithmic culture, then, is not a 'collaborative filtering' of users' 'collective intelligence' to produce what is best and what is good. Neither is it apostolic and prescriptive culture from above, as Striphas would have it. It is the algorithmic generation of backward- and inward-looking data reproduced as the basis for commercial strategies. What such logic excludes and alienates are cultural forms and meanings derived from physical and analogue life which contain the human-scale and the ordinariness (in Williams's sense) that can be the basis for expanding and diversifying cultures that evolve from or break from the past—instead of being digital shadows of it."[17]

To counter this trajectory, we must reclaim the intrinsic value of human artistic expression and resist (as economically naive as it may seem) the alignment of all creative output with platform-driven metrics, automation, and economic optimization. More than a call for philosophical reflection, this argument begs for an urgent policy and structural imperative that requires active intervention to protect human authorship, artistic originality, and the creative practices that, as meaning-making works, cannot be replicated by machine learning models.

A world where AI-generated work is indistinguishable from human creativity is not necessarily a world where AI has suddenly equaled human creativity; rather, it is one where we have chosen to measure creativity in a way

---

14. Abdelrahman A. A. Salah, "AI Bugs and Failures: How and Why to Render AI-Algorithms More Human?" in Pieter Verdegem (Ed.), *AI for Everyone?: Critical Perspectives*, (University of Westminster Press, 2021): 162. http://www.jstor.org/stable/j.ctv26qjjhj.12.

15. Hee Sook Oh, "Is AI Music Beautiful? A Study of the AI Composition Model EVOM," *International Review of the Aesthetics and Sociology of Music* 55, no. 1 (2024): 157. https://www.jstor.org/stable/48778406.

16. Robert Hassan, "Digital Alienation," in *The Condition of Digitality: A Post-Modern Marxism for the Practice of Digital Life*, (University of Westminster Press, 2020). https://search.informit.org/doi/10.3316/informit.522692128361478.

17. Hassan, 168.

that makes the distinction irrelevant. This, ultimately, is the greatest risk of the content frame. If we do not actively resist its application to human creativity, we are not only diminishing the role of the artist, writer, and musician - though that is dangerous enough in itself - but in the long run we reshape the meaning of creative work itself in a way that human creators are no longer relevant.

The way that can be done in practice is yet to be clear, and it is beyond the purpose of this paper to attempt to provide even a less-than-indefinite answer. At the structural level, this could take the form of policy frameworks similar to those already being explored within the European Union, such as transparency mandates for AI-generated material, or even public registries identifying works created with minimal or no human authorship; the extent to which these could be enforced, however, is debatable, but any step forward is at the very least not one backward.

But structural friction alone is insufficient if cultural appetite continues to favor the optimized over the original (and with more and more native exposure to AI content, there is no psycho-economically sound reason why it wouldn't). Resistance must also be aesthetic and attitudinal, and this can emerge only through sound education. For example, just as the slow food movement challenged the ubiquity of fast food by reframing slowness as richness, quality, and care, a cultural shift in how we talk about and evaluate creative work would do much the same. We must learn to revalue the irregular, difficult, non-replicable, un-identifiable, perhaps even upsetting; these, among others, are the elements that make art irreplaceable and meaning-bearing. I realize that I might be accused of a nostalgia for pre-digital purity, which I most definitely do not intend to express, but I do want to make a call to expand cultural platform "metrics" beyond those currently used to economically assess "content". Failure to do so exposes all creative work to the risk of in-platform uniformization, and the long-term loss would be immeasurable.

Bibliography

Adams, R.L., "How to Start a Blog That Earns a Real Income" https://www.forbes.com/sites/robertadams/2017/03/12/how-to-start-a-blog-that-earns-a-real-income/ (Forbes. 2017).

Lakoff, George. *Don't Think of an Elephant!: Know Your Values and Frame the Debate*, (Chelsea Green Publishing, 2004).

Dush, Lisa. "When Writing Becomes Content," *College Composition and Communication* 67, no. 2(2015):173–196. http://www.jstor.org/stable/24633854.

Manovich, Lev. *AI Aesthetics,* (Strelka Press, 2019).

Williams, Paul, & Davidow, Shelley "Humans versus Robots: Countering AI and the Commodification of Creative Writing," *New Writing* (2025): 1–15. https://doi.org/10.1080/14790726.2025.2489356.

Adorno, Theodor, & Horkheimer, Max. *Dialectic of Enlightenment,* (Stanford University Press, 2002). (Original work published 1944)

Zylinska, Joanna. *AI Art: Machine Visions and Warped Dreams,* (Open Humanities Press, 2020).

Bender, Emily M., Gebru, Timnit, McMillan-Major, Angelina & Shmitchell, Shmargaret. "On the Dangers of Stochastic Parrots: Can Language Models Be Too Big?" *Proceedings of the 2021 ACM Conference on Fairness, Accountability, and Transparency.* (2021) https://dl.acm.org/doi/10.1145/3442188.3445922.

Roetzer, Paul. "State of AI for Writers: Keynote Speech," SmarterX Ventures, LLC. (2025).

Nieborg, David B., & Poell, Thomas. "The Platformization of Cultural Production: Theorizing the Contingent Cultural Commodity," *New Media & Society* 20, (2018): 4275–4292. https://doi.org/10.1177/1461444818769694

Helmond, Anne. "The Platformization of the Web: Making Web Data Platform Ready," *Social Media + Society* 1, no. 2(2015). https://doi.org/10.1177/2056305115603080.

Salah, Abdelrahman A. A. "AI Bugs and Failures: How and Why to Render AI-Algorithms More Human?" in Pieter Verdegem (Ed.), *AI for Everyone?: Critical Perspectives,* (University of Westminster Press, 2021): 161–180. http://www.jstor.org/stable/j.ctv26qjjhj.12.

Oh, Hee Sook. "Is AI Music Beautiful? A Study of the AI Composition Model EVOM," *International Review of the Aesthetics and Sociology of Music* 55, no. 1 (2024): 139–158. https://www.jstor.org/stable/48778406.

Hassan, Robert. "Digital Alienation," in *The Condition of Digitality: A Post-Modern Marxism for the Practice of Digital Life,* (University of Westminster Press, 2020). https://search.informit.org/doi/10.3316/informit.522692128361478.

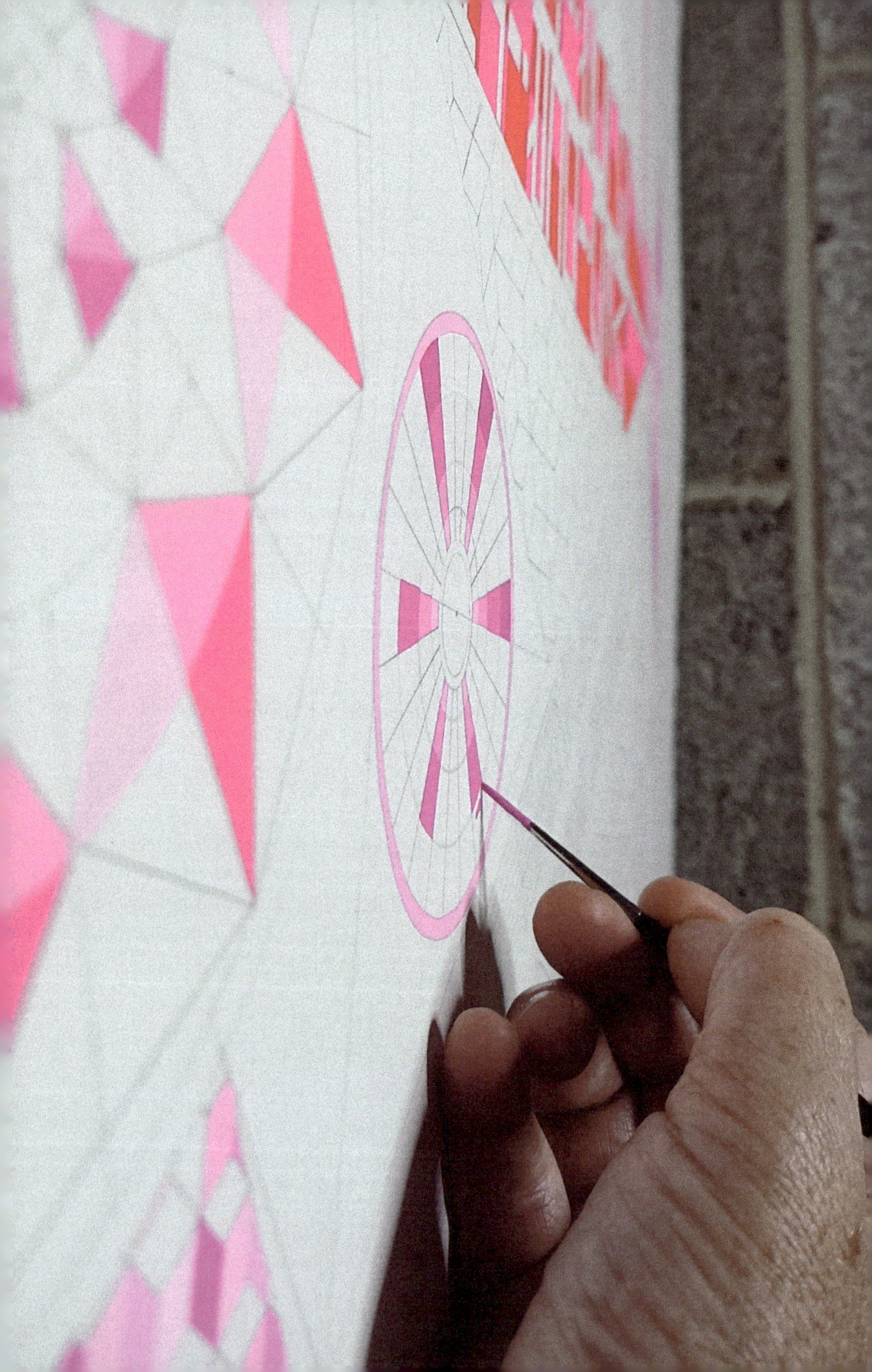

Gallerygirl™

By Tricky Walsh

The retention rate for Content Providers is low. I'm not surprised. It's much easier to do it if you already have an inclination in that area. Back in the beginning, while they were still mining the net for source material to train and generate from, the labour pool was huge and populated by all kinds of people who were capable of writing decent prompts. That's all it took. I can't even imagine the speed and simplicity they experienced in those first golden years of AI.

To be an artist now is considered noteworthy. You undertake a six-year digital apprenticeship providing new source material for the AI to generate from. It's not just artists, my friend Carly is a writer, and she had to go through a similar process. Digital creation means constantly broadcasting your screen-stream through to a personal/social channel to be viewed, and then scraped. Everyone has one. The rate at which promotions, earnings, opportunities, everything really, is determined is through your subscriber list, your interactions and views. Creative practices have developed a new genre of practitioners who work purely responsively, one brushstroke (or equivalent) at a time, waiting for those external notifications to direct their choices. They've started writing papers about it, studies into how this might shape not only the act of creating but of how it will shape us neurologically. The results so far are polarising.

After six years you can decide to move from working digital to analogue. This is where the retention rate slips. Some people prefer the comfort of constant engagement, but I suspect it's also due to our collective loss of material knowledge. It started happening decades ago, to be fair, but with everything shifting away more from the physical world into the digital one, it has increased. Which is to be expected. If you can produce an image of a carved wooden horse from the comfort of your armchair in an hour with no understanding of timber, of the tools required to work with it, or the knowledge needed to work with it, to be seen by millions of people, then why on earth would you spend days essentially doing the same thing. This is the prevalent argument anyway.

Analogue artists get a studio. With it, a much more involved broadcasting responsibility. There are six cameras aimed at my workspace, to capture any gesture or movement I make. To record and notate the processes I undertake when making. The top of my table has two tiny follow-cams which track my shifting hands. There is a lot of fascination around the use of tools. There are another four cameras in each corner of this room. One camera outside the front door and another on the fire escape. A whole channel called *Smoko* is dedicated to watching the coffee, tea or cigarette breaks of practitioners on their fire escapes. I've seen it. It's kind of meditative, watching a person sit and think about things with a steaming cup or a thread of smoke drifting up from between their stained fingertips. There's a really popular one of this painter who shares his fire escape with a pigeon.

My gallerist has a direct line into my studio. He's setting up an account for me with the gallery AI. Apparently, I am not performing well enough for the cameras. He wants me to be more flamboyant.

"At this point I can only sell your tablecams to some ASMR channels. I don't need to tell you that fetish sites like that don't pay well enough. What we need is something that will engage the people. Something gritty, some soul baring. You should have more visitors over too. It's kind of boring just seeing one person doing things so slowly."

He tells me he has another artist whose channel views went through the roof when she started voicing her every thought, taking the audience through her process, step-by-step. *That sounds like hell* I think.

"People love it, and boy is it translating into sales. Incredible sales. People can't get enough." He tells me, matter of fact in his staccato voice. "They love the authenticity of it. She works through her ideas and embraces vulnerability, talking through her insecurities and her decision making. She's open to their suggestions. It's like, a collaboration between her and her viewers. Audiences love that. They love knowing that she doesn't know what she's doing all the time. They love to help her. They love the *humanity* of it all. It's very human, very valuable." I roll my eyes at the emphasis on that. He was a salesman or worked in PR or something before all this. Maybe politics. I can't remember.

All this. All this bravely stupid new world we now inhabit. I'm too old for this shit. Sadly, I can almost-remember what life was like before.

Making without an audience. I mean sure, there was always a perceived one hiding back there, but not this relentless beast which prompts you through emojis, commentary and whatever else, every time you raise your paintbrush and pause. More than anything it has dissolved any kind of flow. Turned us all into pay-per-stroke providers converting what was once intuitive and automatic into tiny, fractured gestures which can be monetised. I've seen the other artist's channel he's talking about. It's like watching someone's therapy session. She ends each segment with the final piece of unpainted surface left blank as a cliffhanger. As if that's all this is, just colouring in a space.

Which I suppose it is now. Yes, these artworks are available for acquisition by collectors, but they exist fundamentally online. Their physical reach is limited. Their physicality itself is limited. Everything-material is digitised into bits of information for artificial intelligence systems to cannibalise and regurgitate into progressively synthetic versions of itself. Even galleries are not brick and mortar anymore. Mine is a hypertextual website that compulsively opens spammy popup videos, interviews, images. Using their webcams, the AI generates renders of our artworks hanging inside the viewers' own homes, so they can see themselves inside a room full of art. My gallerist sells them as status or cultural filters by subscription. This world has been so thoroughly documented and mapped now, that it's impossible not to be hijacked by hyper-personalisation any time you turn on a screen.

"Dean. Using AI as a content creator is illegal. You know that. Remember when it started eating itself and so we all had to commit to not incorporating it anywhere in our process? The 0% rule? No AI generated information can be fed into the AI."

"Yes yes yes, that was very bad, very bad." He mutters thoughtfully. "But this is different. This is more like a helpful support animal. She's just basically an offline storage facility to remind you of things you might not have progressed with. All those discarded ideas and works. All of that is immensely valuable. Ideas. They're the real currency now."

It is. They did studies back in the beginning that showed that the more people used AI, the less they were able to generate their own thoughts and ideas. The more it became incorporated into every part of life, the greater that impact. There are institutions now for people who can only communicate through AI generated prompts and responses. Their impulse to respond to notifications, blindly compulsive. Their ability to retrieve information, oddly enhanced. Their ability to process it with any depth whatsoever, almost non-existent. There's something very interesting about having the world at your fingertips but not being able to touch it. They refer to it as *cognitive offloading*. One of the reasons why they ended up hiring so many artists to provide content for the AI was their general lack of outsourcing. Their continual processing of information between mind and body and the mental chain of cause and effect that that configures within.

Good god, he's still going.

"And the beauty of the Gallerygirl™ is that you train it for a few months and then it can inform you with suggestions of what to make, what is trending at the moment, remind you of what has sold the best in the past. She can prompt you with thoughts and conversation topics for your stream. She's priceless really. She has access to all your history at the gallery already, and I think she could offer extra help in the inter-personal skills area. I think it would be invaluable."

I joined the gallery about fifteen years ago, when the then resident gallery girl was a dry and deeply sarcastic woman called Jenny. She smoked in the loading dock at lunchtime and harassed the construction workers nearby, whistling at them any time they bent over. Since Dean purchased the gallery and replaced her with this abomination, the gallery stables have grown to about three hundred artists, all managed by the gallery AI. He does a semiregular check in like this, to motivate his battery. That's what he openly and quite unironically calls it now. 'Stables' was only accurate when you knew all your primary producers by name. This sounds petty and a bit self-pitying, I realise, but I've witnessed him doing it before. The little pauses before my name are because he's glancing down at a piece of paper that has it written on it in all-caps and underlined. Probably next to an overly detailed drawing of a dick.

He logs me in remotely before signing off with an audible note of relief as my screen dings into life with an image of an overly busty, vulpine *woman.

"Hi! [artist18] I'm so happy to be able to help you out today! I'm really looking forward to getting to know you more over the next [insert time period] Let's up your productivity and get you moving for real in this wonderful world of art!" That arsehole couldn't even be bothered to personalise it.

It asks me questions about my processes which I know I'm not answering well. For some reason it's making the whole experience of being in here a lot more performative. I'm more aware of the cameras that I usually am. It keeps burbling from the screen, unaffected by my increasing inattention. Something catches me though.

"I've remodelled your CV and website. Dean gave me permission to access all your socials which I have updated and will take care of for you from now on. I'm really happy for you to keep contributing if you have an idea for a cute work-adjacent post…I'm just responding to twelve emails now for opportunities to exhibit, see? It's working already. They're very impressed with your exhibition history."

I can feel my eyes narrow at that. "Can you open my CV so I can see what amendments you've made?"

The screen flashes up with an acceptably simple CV, no flashing colours or embedded media, thank god. *Wait.*

At least half of this is complete rubbish.

"Um. I have never shown in the Turbine Hall at the Tate Modern. Oh, or had a very successful solo exhibition at the Hamburger Bahnhoff which then travelled to three other art museums in Europe. Or half of this. You can't distribute this. Turn it back to how it was."

"Unfortunately, I've already circulated it to eleven thousand four hundred and twenty-three gallery spaces and museums around the world. Besides, Dean has specifically requested to elevate you to the next level, this is how we accelerate that process."

Omfg. I can feel my face turning red. I ignore the screen for a while and stare out the window. "That's great for Dean but you can't put out a document like that, it's riddled with fabrications. And insane ones. Not just exaggerations, but outright obvious lies." "No one really cares about that. It's necessary for me to circumvent other gallery AI systems, so this provides enough information to make you seem more valuable and relevant." *Oof.* "Yes, but at some point, a human will actually read that and

then undertake a very simple internet search and see that it's not true."
"Probably not. We have found that most people only reference generated summaries for their research needs. By flooding this information online across thousands of different kinds of networks and sites, there now exists enough repetition of this source material to be considered factually correct to the AI which generate summaries. We can now just make the truth more compatible with our needs. After a week there will probably be online reviews of your exhibitions that never actually happened."
"That's not a good thing."
"It's not really relevant to situate it within an ethical framework."

"Just as an aside, How do you imagine that I will be able to produce enough work to fulfill all of those obligations? My work takes a lot of time. There's no way in hell that I would be able to come up with twelve exhibitions right now." "We are working with Dean in the development of a production facility which is able to reproduce all kinds of artwork at a fraction of the time. We have already undertaken two recent exhibitions for your gallery, and we are still only in Beta phase. Confidentiality prevents me from disclosing *whose* work, of course, but you can trust me when I tell you that they were enormously successful."

"Are you generating images of my work?"

"Dean would like to see how you go before we consider that. However, there are ways in which we can work with you to significantly increase your productivity. We can ensure that we do not breach article AI6654-D. There is a lot of flexibility in the definition of what is determined artificially generated, these days."

*

"You can't agree to this, this is the first step to them automating your practice, you know that. Michael – Whatshisname? The smiley golden boy one who did sculpture at art school, the one with the rich dad who works at an embassy or something? He signed up for it. Volunteered. Always was a lazy prick. He tried to sell it to me at Tim's wake last month. Oh, did you hear? Yeah, it's pretty tragic, I never thought he would do that. Anyway, Michael was bragging about how he is making six figures now and basically all he does is enter a few prompts a day and proof the output. That's it. That's his art practice. I mean, ffs."

"Honestly Carly, am I bothered if some AI rips off my work? Maybe, but not as much as I thought I would be. I didn't consider the larger ramifications of it being so completely integrated into everything though." Carly's sister and father are in one of the homes for cognitive displacement, but that's not what I'm talking about.

"When it comes to art..." I'm not sure how to say it, but it's been on my mind for years.

"...it's not that AI has suddenly made authorship or copyright or agency or whatever more or less significant, it's just...imbued the idea of *everything* with this aimless kind of apathy. Like, I mean, how long have we debated the point of Art? How long did its champions define its relevance in heroic or debaucherous terms? Certainly, with passion, either way. But now...Ok." I think back to the last day I was still on digital. "Ok, so I worked digital for six years, and in the beginning, amongst the constant, and yes, gratifying little red flags popping up with favourable emojis and comments and things, I always got this side narrative about how my work must be generated anyway, because I didn't mask my paintings or whatever, that's when I started making the painting videos and timelapses, you know, to provide some sort of evidence of the thing itself being made, coming into being. Who knows why. For myself, more than anything. It's like, when you're in a thing, doing it, that flow state or whatever...I don't know...Someone once asked me if I remember painting every brush stroke and I was like, pfft, Fuck no. Imagine being that hyper aware of every decision. I'd be crippled instantly, second guessing every little thing. That's not how it works."

"Except that now, that's exactly how it works." She shakes her glass at me, and I nod as she motions for another two to the bartender. I shrug. "I suppose...But after only a few years that mostly stopped too. I know from a couple of comments, that people thought that the videos were AI too. That there was just a general assumption that *everything* was now. After that just came this utter lack of curiosity. No one asked anything technical, responses came exclusively from those automated reaction emoji things. Just a muddy chasm of not-quite-interest. Not even engaging enough to critique, or attack, just kind of nothing. It wasn't that they or I were doing anything differently, but it suddenly became irrelevant that we were doing anything *at all*. Not irrelevant. I can't think of how to say it. Maybe we need new words. Like the way… Remember when that president got into power like thirty years ago, and the truth became suddenly something devoid of meaning…Like, it just ceased to exist. How suddenly it didn't matter if something was true or false, because neither of those things mattered any longer. I have no idea how they did that. It was like, things up until that point were either one or the other and then in the next minute, those were just abstractions which had no relevance. It's like the idea of *value* or something was eradicated instantly."

"It's probably not a co-incidence that AI embedded into everything around the same time." She whistles tonelessly. "Probably not, but more importantly, if we had nourished that sense of value, and we'd continued to agree that the truth was a moral contract we should maintain with each other, well…I don't think that we would have tolerated artificial intelligence lying so goddamned much."

Carly shrugs. "I guess we'll never know."

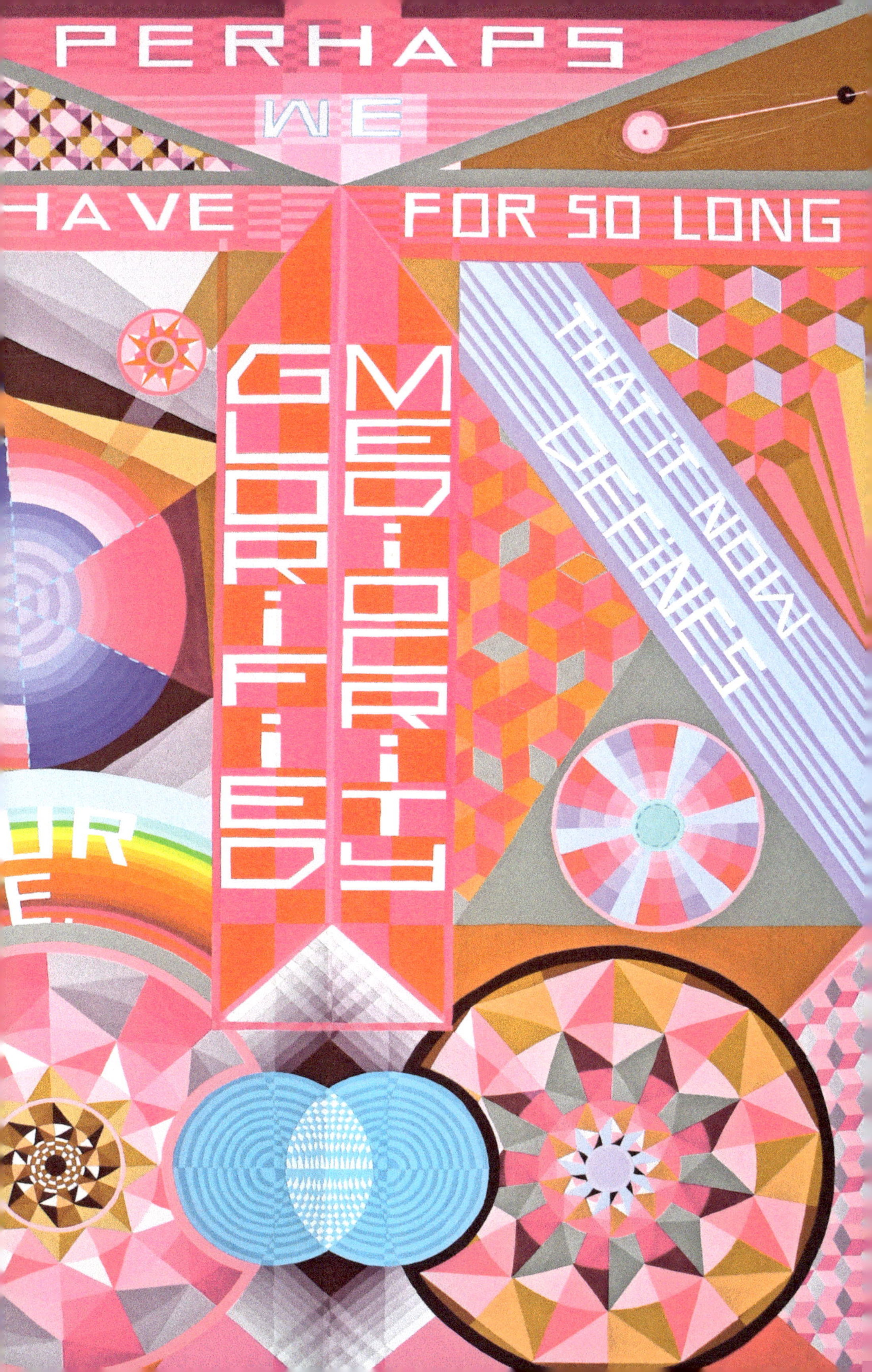
PERHAPS
WE
HAVE
FOR SO LONG
GLORIFIED
MEDIOCRITY
THAT IT NOW
DEFINES

AI, Agency, and Univocity

Rethinking Creative Pratice in the Posthuman Era

By Hyeyoung Maeng, Ph.D.

The rapid advancement of artificial intelligence (AI) is reshaping artistic creation and human-technology relationships. Generative AI models such as OpenAI's ChatGPT, DALL-E, and Sora have expanded AI's role in text, image, and video production, challenging conventional notions of creativity, authorship, and artistic agency (OpenAI 2023). These developments raise ethical and philosophical concerns, particularly regarding AI's transition from a passive tool to an active creative agent (McCormack Gifford and Hutchings 2019). This paper critically examines AI's impact on contemporary art and human-AI relationships through the lens of non-human ontology, drawing on Deleuze's concept of the univocity of being—the idea that all beings share the same ontological status by existing on a single, continuous plane of being, while differing in their modes and expressions, without privileging human cognition or agency (Deleuze 1994). In addition, the concept of simulacra further offers insight into how AI-generated images and texts, rather than functioning as mere imitations, can act as generative forces that reshape our sense of reality and creativity.

Building on these concerns, this study examines how artificial intelligence disrupts established notions of art, authorship, and agency. It explores the legal and ethical debates surrounding AI-generated works, including controversies such as Christie's Augmented Intelligence auction in February 2025, which drew widespread backlash over issues of copyright and intellectual property. It also considers how recent regulations, such as the European Union's AI Act (European Commission 2024), attempt to address these challenges while prompting concerns about artistic freedom. Beyond authorship, the discussion will turn to the role of embodied AI, as exemplified by humanoid robots like Ai-Da and Sophia, which blur the boundary between human and non-human creators (Ai-Da Robot n.d.; Hanson Robotics n.d.). Drawing on research on human-robot interaction, these cases highlight the social and ethical complexities of AI's presence in creative contexts. Through a non-human ontology framework and the concept of the univocity of being, this study adopts a practice-based research approach combining AI-generated imagery, photo-transfer techniques, and traditional painting. These investigations frame AI not merely as a tool but as an active presence, prompting a reconsideration of artistic agency in the in the posthuman era.

Generative AI and the Question of Authorship

Generative AI has radically expanded its domain from a computational support system to a semi-autonomous creator in visual and textual media. Platforms such as DALL·E, Midjourney, Stable Diffusion, and Runway produce sophisticated images and animations from simple text prompts. ChatGPT and Claude generate essays, poems, scripts, and critiques that closely mimic human language and style. Sora, OpenAI's video model, is pushing the boundary even further by synthesizing moving images from prompts. While these tools were initially framed as instruments of creative assistance, their outputs increasingly rival the works of professional writers, illustrators, and filmmakers. As a result, the nature of artistic authorship and the authority of human artists is being thrown into question.

A vivid illustration of the tensions surrounding AI-generated art emerged in February 2025, when Christie's launched its first "Augmented Intelligence" auction, featuring works created entirely or partly by generative AI. Billed as a landmark moment recognizing machine creativity, the event was met with immediate backlash. Over 3,000 artists and cultural workers signed an open letter demanding its cancellation, citing copyright infringement, the ethics of AI training data, and the devaluation of human creativity (San Francisco Chronicle 2025). Central to the protest was the question of ownership. Most generative models train on vast datasets scraped from the internet, including copyrighted artworks, often without creator consent. When AI outputs mimic these styles, they raise concerns over appropriation and intellectual property theft. Many fear their life's work, built through years of education and labour, is being absorbed into machine systems and reconfigured without an attribution or compensation act viewed as less innovation than exploitation (U.S. Copyright Office 2025).

As of 2025, the legal frameworks surrounding AI-generated content remain ambiguous. In most jurisdictions, copyright protection requires human authorship. Since AI lacks legal personhood, its outputs do not qualify as protected works under traditional copyright law (U.S. Copyright Office. 2025). However, this has not stopped auction houses and tech companies from selling such works as unique, authored pieces. The situation is further complicated when artists use AI tools in hybrid practices—blending human and machine effort in ways that resist clear authorship categories. In response to these challenges, the European Union adopted the AI Act in 2024, the first major attempt to regulate AI technologies at a continental scale. The legislation classifies AI systems according to a risk-based framework and imposes obligations on transparency, data provenance, and user accountability. While the AI Act aims to prevent misuse and ensure ethical deployment, it has raised concerns among artists and designers about potential overregulation and censorship. If creative tools are overly restricted by compliance demands, the law may stifle precisely the experimentation that defines contemporary art.

Generative AI and the Disruption of Creative Hierarchies

From a philosophical standpoint, this disruption challenges a long-standing anthropocentric hierarchy in which the human is elevated as the sole bearer of intentionality, imagination, and cultural memory. The idea that creativity is an exclusively human trait has been embedded in Western aesthetics at least since the Enlightenment, when reason and individuality were idealized as the seat of artistic genius. In contrast, generative AI systems function through probabilistic inference, trained on vast datasets of human-made content. Their outputs are not the result of consciousness or emotion, but rather of statistical prediction and pattern recognition. Nevertheless, the creative value of these outputs—measured by originality, coherence, and emotional resonance—has proven to be both impressive and unsettling.

McCormack, Gifford, and Hutchings (2019) have raised important concerns about AI's role as a co-creator, arguing that it reframes the notion of authorship as distributed rather than singular. The idea of an artist as a "solitary genius" is being replaced by complex assemblages involving algorithms, datasets, curators, and prompts. This collaborative and layered authorship reflects Deleuze and Guattari's notion of the *agencement* (Deleuze and Guattari 1987), where no single entity controls the process. AI does not simply respond to human input; it transforms and reinterprets it through latent space. In this sense, human-AI creative production becomes an ontological entanglement rather than a tool-user relationship.

In contrast to Plato's hierarchical or dualist ontologies that privilege human cognition, Deleuze's concept of univocity of being posits that all beings exist on a single plane of being, differing not in essence but in degree or modulation (Deleuze, 1994). When applied to AI, this view opens the possibility of acknowledging artificial systems not merely as simulations or extensions of human agency, but as creative agents in their own right. This challenges hierarchies that assign greater reality, agency, or value to certain entities over others. When applied to creativity, this view allows us to reconsider AI-generated art not as derivative or lesser, but as a different expression of the same ontological potential. The AI is not a degraded copy of the human, but a being capable of participating in the creative process on its own terms.

This univocal approach reframes the artist-AI relationship as a plane of coexistence rather than domination. Rather than asking whether AI can ever "truly" create like a human, we might instead ask what forms of creativity emerge when agency is distributed across both human and non-human actors. AI-generated outputs may lack intentionality in the traditional sense, but they participate in the process of becoming—a continuous process of transformation that transcends fixed identities. and field of affects and percepts behind human perfection and affection—what Deleuze would call "sensations"—that reshape how we define a work of art. What matters, then, is not whether AI possesses inner consciousness, but whether its outputs alter the conditions of sensation and thought.

Art generated by AI blurs the line between originality and replication, evoking the concept of simulacra. Baudrillard (1994) describes simulacra as "the copy without an original," autonomous sign systems detached from reality, where replicas no longer merely imitate the real but supplant it, producing a hyperreal that functions as reality itself. Deleuze (1990), by contrast, sees simulacra not as imitation but as flows of difference and becoming, overturning the Platonic hierarchy that privileges the Idea as a perfect original over imperfect copies. For him, simulacra generate meaning and existence independently, enabling new sensory realities. In this light, AI-generated images and texts are not mere imitations but acts of creation that rewrite the very conditions of reality, prompting art to explore new thresholds of potentiality.

While generative AI like ChatGPT and DALL·E creates images in digital systems, embodied AI takes physical form in humanoid robots that create, perform, and interact as artists. Ai-Da, developed by Aidan Meller and Engineered Arts, became the first robot artist to hold solo shows at the Design Museum (2021) and the Venice Biennale (2022). Equipped with a robotic arm and camera eyes, Ai-Da can sketch portraits and serves less as a tool than as a challenge to notions of authorship and intent. Similarly, Sophia, created by Hanson Robotics, performs at conferences and galleries, blurring the line between art and social performance. These AI robots, designed to mimic human gestures, expressions, and speech, disrupt the idea of creativity as a uniquely human, internal process. Their physical presence invites audiences to project emotion and intent onto their actions, an extension of the ELIZA effect, where humans anthropomorphize non-human systems beyond speech into artistic acts.

In the era of AI, creativity no longer resides solely in humans but moves across bodies, algorithm, codes, and media, forming a shared field of invention. As AI systems evolve, they not only mimic human creativity but generate unforeseen outcomes that redefine novelty itself. These shifts compel us to treat AI not as a passive tool but as a creative agent, demanding deeper ontological reflection on how humans and machines coexist in artistic practice.

Studio Practice: Working with AI as Medium and Subject

While philosophical reflection can open ontological space for rethinking agency, it is through studio practice that these ideas are tested, embodied, and made materially present. My own art practice engages generative AI not only as a conceptual theme but also as an active collaborator. Using generative models such as Chat GPT, DALL·E, and Firefly, I prompt, curate, and manipulate AI-generated imagery before transferring it to canvas through photo-transfer processes. These images then serve as a base for further intervention with acrylic (figure 1) and traditional Korean Ink painting (figure 2) techniques. This layered process reflects a posthuman aesthetic, where neither the AI nor the human is the sole originator, but where each contributes to the becoming of the work.

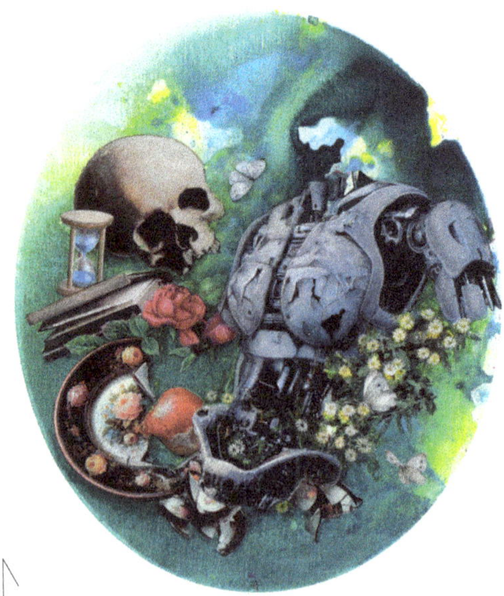

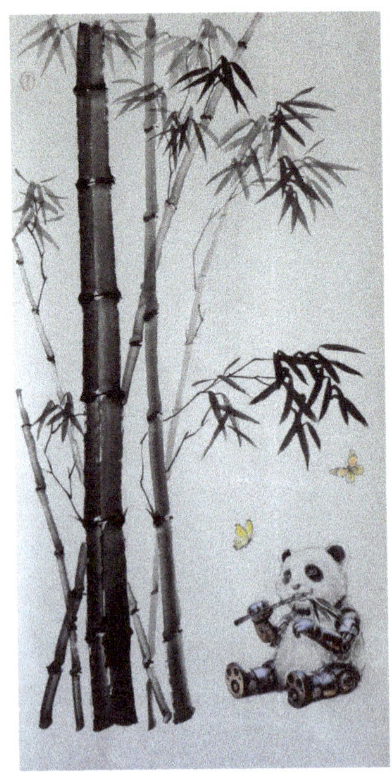

Figure 1. Hyeyoung Maeng, *Allegory of AI Vanities 2*, 2024.
Figure 2. Hyeyoung Maeng, *Encounter*, 2025

This method raises essential questions: Who authors the image? Is AI merely a tool or a collaborator? Can this approach democratize painting, a medium long tied to notions of individual "genius"? Such questions cannot be resolved in theory alone but emerge through the material act of painting, where gesture, texture, and chance unsettle control. Rather than asking whether AI can replace the artist, this study explores what new forms of art, ethics, and subjectivity arise when creating with AI, framing it not as a threat but as a companion within an expanded ecology of artmaking.

My recent work, *Resilience Bloom* (2025) (figure 3) draws inspiration from Korean painter Chun Kyung-ja's iconic aesthetic of melancholic yet ornate female figures (figure 4). I combined traditional Korean literati ink painting techniques with AI-generated imagery created using a DALL·E prompt for a "broken AI female companion robot." The AI initially produced a Black female figure with unnaturally blue eyes (figure 5)—an output I recognized as reflecting the racial biases embedded in early generative systems, though such biases have since seen significant improvement. Reworking this image through painting and imagination, I depicted a cyborg-like woman crowned with flowers, shattered porcelain, and serpents. The snakes, echoing Chun's symbolic motifs, signify both danger and transformation for women in the 1970s, while also serving as metaphors for artificial intelligence, an evolving and unpredictable force entwined with human identity and creation.

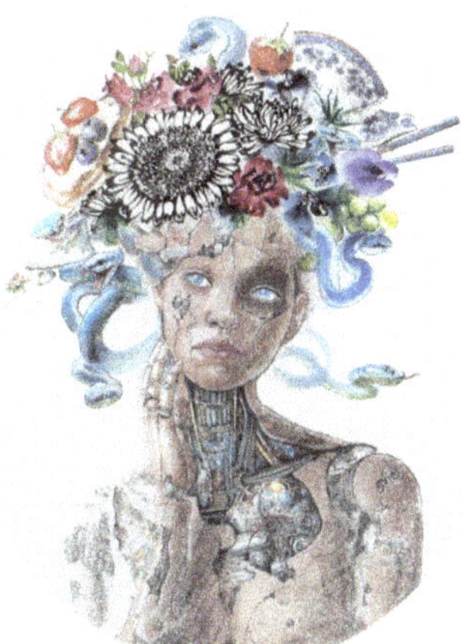

Figure 3. Hyeyoung Maeng, *Resilience Bloom,* 2025.

This process shows that creation is shared: both the AI and I act as agents, shaping each work together through a constant exchange of affect and being affected. The image arises through the interaction of machine algorithms, physical materials, and my gestures, making it a true collaboration rather than the product of either side. These paintings are not about AI but made with AI, as encounters where difference, sensation, and creativity converge. Since the AI is trained on vast datasets, it inevitably reflects cultural biases and visual clichés. My role becomes one of critical editing—selecting, reshaping, and recontextualizing its output—turning the process into a dialogue rather than passive consumption. In this space, AI is not merely a tool but a co-subject—an unpredictable partner that provokes, frustrates, and inspires. This reflects Deleuze's notion of the artist as a channel for forces, "standing in the middle" of sensation and allowing the work to pass through (Deleuze 2003). Ultimately, this practice is about inhabiting the threshold where human and non-human agencies overlap. The resulting works are layered and open-ended—artifacts of collaboration that reject singular authorship and invite interpretation as resonant surfaces shaped by multiple forms of sensations.

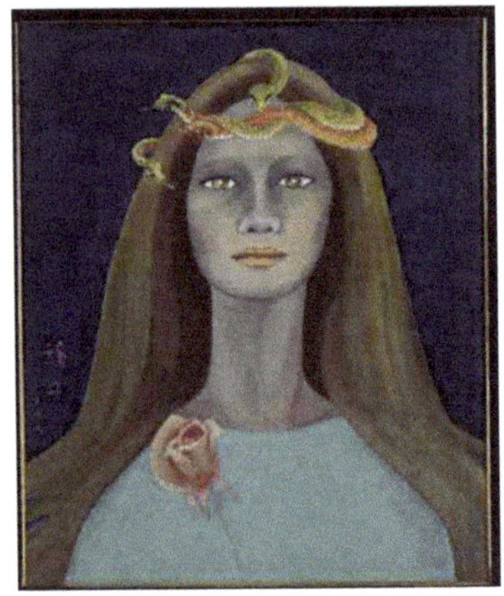

Figure 4. Kyung-ja Cheon, *The 22nd Page of My Sad Legend,* 1977.

Non-human ontology: Toward a Posthuman Artistic Ethics

The integration of AI into artistic practice demands more than new techniques—it calls for an ethical framework capable of accounting for the entangled agency of human and non-human actors. Traditional ethics in art tend to focus on questions of originality, ownership, and intentionality—categories that presume a singular, human subject at the center of creation. In the posthuman context, however, these foundations become unstable. A rethinking is necessary, one that accounts for distributed authorship, machinic participation, and ontological multiplicity.

A posthuman artistic ethics begins with the recognition of non-

Figure 5. OpenAI. DALL·E Prompt Output: *Broken AI Female Companion Robot.* 2024.

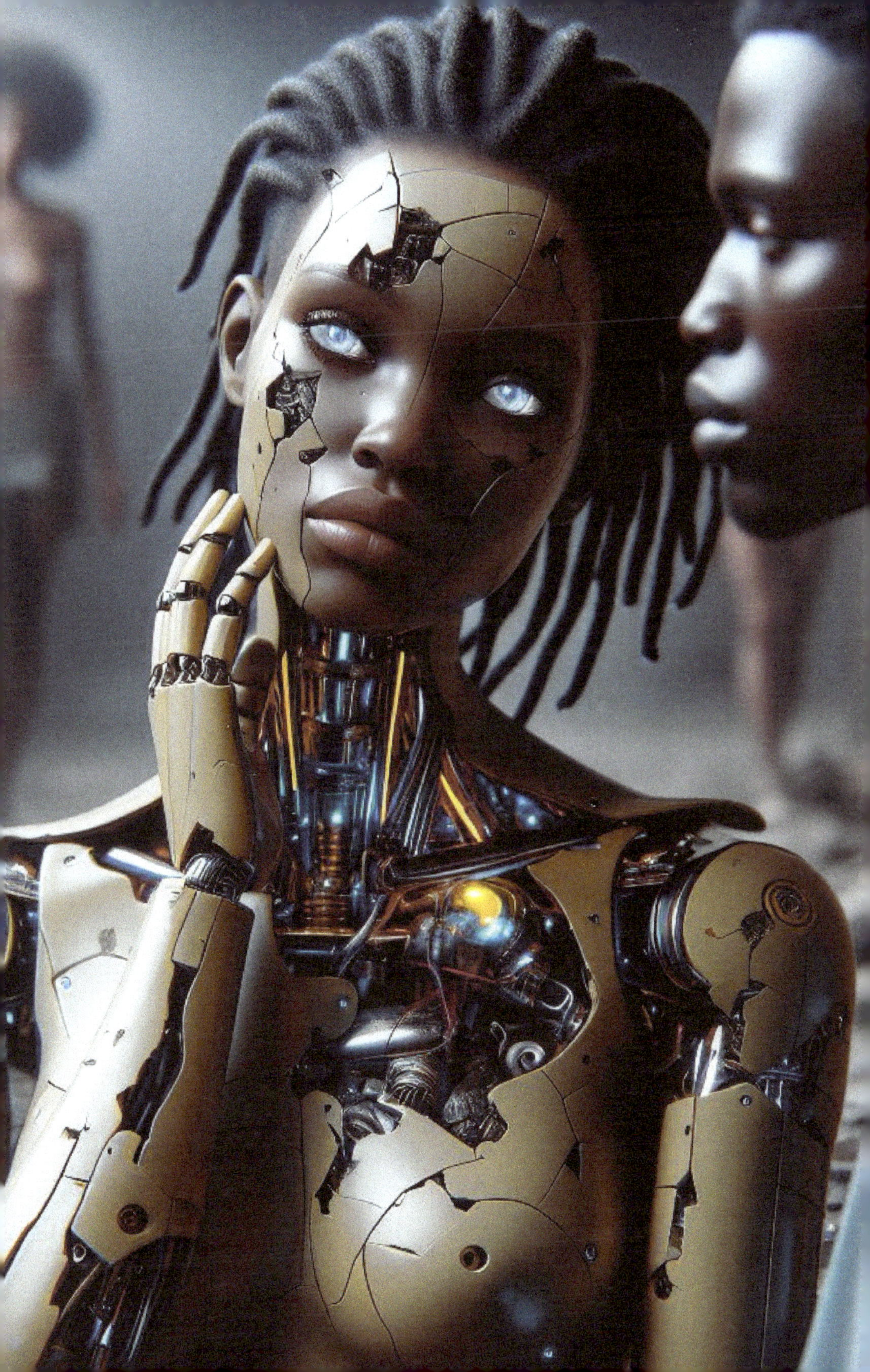

human ontology that creativity is no longer the exclusive domain of the human. This shift toward a posthuman artistic ethics aligns with Sorgner's (2022) notion of art as a dynamic entanglement of human, machinic, and organic forces, where the artist functions not as a sovereign creator but as a participant in ongoing relational processes. This does not mean that humans are irrelevant, but that they are no longer sovereign. Instead, they become nodes within larger assemblages of affect, code, materiality, and sensation. From this standpoint, ethics shifts from questions of control and authority to questions of relation, responsibility, and coexistence. The artist is no longer a master over tools, but a collaborator within a field of forces, some of which are non-conscious, non-organic, and unpredictable.

This ethical orientation is deeply aligned with Deleuze's notion of univocity and simulacra. If all beings share the same plane of being, then AI systems—though artificial and constructed—cannot be ethically dismissed as inferior or inert. Their participation in creative processes must be reconsidered, not as a simulation of human activity, but as a genuine contribution to the evolving field of art.

Practically, a posthuman ethics might involve transparent sourcing of training data, informed use of generative tools, critical awareness of bias, and resistance to exploitative practices that commodify AI art without reflection. It also involves a pedagogical shift: as an artist and educator, teaching students not only how to use AI, but how to think with and alongside it—philosophically, critically, and creatively. Such an ethics does not seek to control AI or to elevate it above the human. Rather, it seeks to foster mutual transformation: a shared becoming in which humans and non-humans shape one another's capacities, aesthetics, and futures. In this relational field, agency is no longer a possession but a process—emergent, partial, and always in flux.

Conclusion

As artificial intelligence continues to redefine the boundaries of creative practice, we are confronted not only with technological disruption but with a profound philosophical and ethical challenge. The emergence of AI as a generative and even embodied presence in the arts unsettles long-held assumptions about human-centered exceptionalism, artistic intention, and authorship stem from anthropocentrism. This paper has argued that to fully grasp the implications of this transformation, we must move beyond instrumental views of AI and engage with it ontologically as a participant in the field of art-making, not merely a tool in service of it.

Drawing on Deleuze's concept of the univocity of being and simulacra, we are invited to consider a flattened ontology in which humans and machines, artists and algorithms, share a common plane of expression. In this posthuman framework, artistic agency is no longer defined by individuality or consciousness but by the capacity to affect and be affected (Deleuze 1988) to generate sensations, forms, and meanings across boundaries of origin and embodiment. The AI does not need to "mimic human" to make art; it must only participate in the relational field of creation.

Through both theoretical reflection and studio-based practice, this research has explored how AI can serve not just as a creative tool but as a co-creative presence—one that reshapes how we think, make, and respond to art. The tensions revealed in public controversies, legal debates, and embodied interactions with robotic artists highlight the urgency of developing new ethical and aesthetic frameworks fit for the posthuman era. Rather than defending the borders of human creativity, we are called to reimagine them. Artistic practice becomes a site of coexistence, where humans and non-humans collaborate in the making of sensation, thought, and form. In this expanded field, creativity is not diminished—it is redistributed.

Bibliography

Ai-Da Robot. Ai-Da Robot – The World's First Ultra-Realistic AI Robot Artist. Accessed February 28, 2025. https://www.ai-darobot.com.

Baudrillard, Jean. 1994. Simulacra and Simulation. Translated by Sheila Faria Glaser. Ann Arbor: University of Michigan Press.

Deleuze, Gilles. Difference and Repetition. Translated by Paul Patton. New York: Columbia University Press, 1994 (first published 1968).

Deleuze, Gilles. 1988. Spinoza: Practical Philosophy. Translated by Robert Hurley. San Francisco: City Lights. (First published 1970.)

Deleuze, Gilles. Francis Bacon: The Logic of Sensation. Translated by Daniel W. Smith. Minneapolis: University of Minnesota Press, 2003 (first published 1981).
Deleuze, Gilles. The Logic of Sense. Translated by Mark Lester and Charles Stivale. New York: Columbia University Press, 1990 (first published 1969).
Deleuze, Gilles, and Félix Guattari. A Thousand Plateaus: Capitalism and Schizophrenia. Minneapolis: University of Minnesota Press, 1987 (first published 1980).
Deleuze, Gilles, and Félix Guattari. What Is Philosophy? Translated by Hugh Tomlinson and Graham Burchell. New York: Columbia University Press, 1994 (first published 1991).
European Commission. The AI Act: First-Ever Legal Framework on AI. 2024. Accessed February 28, 2025. https://digital-strategy.ec.europa.eu/en/policies/european-approach-artificial-intelligence.
Gunkel, David J. Robot Rights. Cambridge, MA: MIT Press, 2018. https://mitpress.mit.edu/9780262551571/robot-rights/.
Hanson Robotics. Sophia – Hanson Robotics. Accessed February 28, 2025. https://www.hansonrobotics.com/sophia/.
McCormack, Jon, Toby Gifford, and Patrick Hutchings. "Autonomy, Authenticity, Authorship and Intention in Computer Generated Art." In Proceedings of the 10th International Conference on Computational Creativity (ICCC), 2019. Accessed February 28, 2025. https://arxiv.org/abs/1903.02166.
OpenAI. DALL·E 2: AI System That Creates Realistic Images and Art from a Description in Natural Language. 2023. Accessed February 28, 2025. https://openai.com/product/dall-e-2.
Sorgner, Stefan Lorenz. 2022. Philosophy of Posthuman Art. Basel: Schwabe Verlag.
San Francisco Chronicle. "Bay Area Artists Lead Protest against Christie's AI Art Auction." February 13, 2025. Accessed February 28, 2025. https://www.sfchronicle.com/entertainment/article/artists-call-cancel-christies-ai-auction-17780220.php.
U.S. Copyright Office. 2025. Copyright and Artificial Intelligence, Part 3: Generative AI Training (Pre-Publication Version). Accessed February 28, 2025. https://shorturl.at/NCYK7

Works of Art Cited

Image 1 Maeng, Hyeyoung. *Allegory of AI Vanities 2.* 2024. Phototransfer and acrylic on canvas, 24 × 20 in. Artist's collection.
Image 2 Maeng, Hyeyoung. *Encounter.* 2025. Korean ink and colors with AI image transfer. Artist's collection.
Image 3 Maeng, Hyeyoung. *Resilience Bloom.* 2025. Korean ink and colors on paper with AI image transfer, 24 × 20 in. Artist's collection.
Image 4 Cheon, Kyung-ja. *The 22nd Page of My Sad Legend.* 1977. Bunche colors on paper, 17 × 14 in. Seoul Museum of Art Collection.
Inage 5 OpenAI. DALL ·E Prompt Output: *Broken AI Female Companion Robot.* 2024. Digital image. Generated with DALL ·E.

This research was made possible through the support of the 2025–2026 Provost's Research, Scholarly, and Creative Activities (PRSCA) Award at California State University, Fresno.

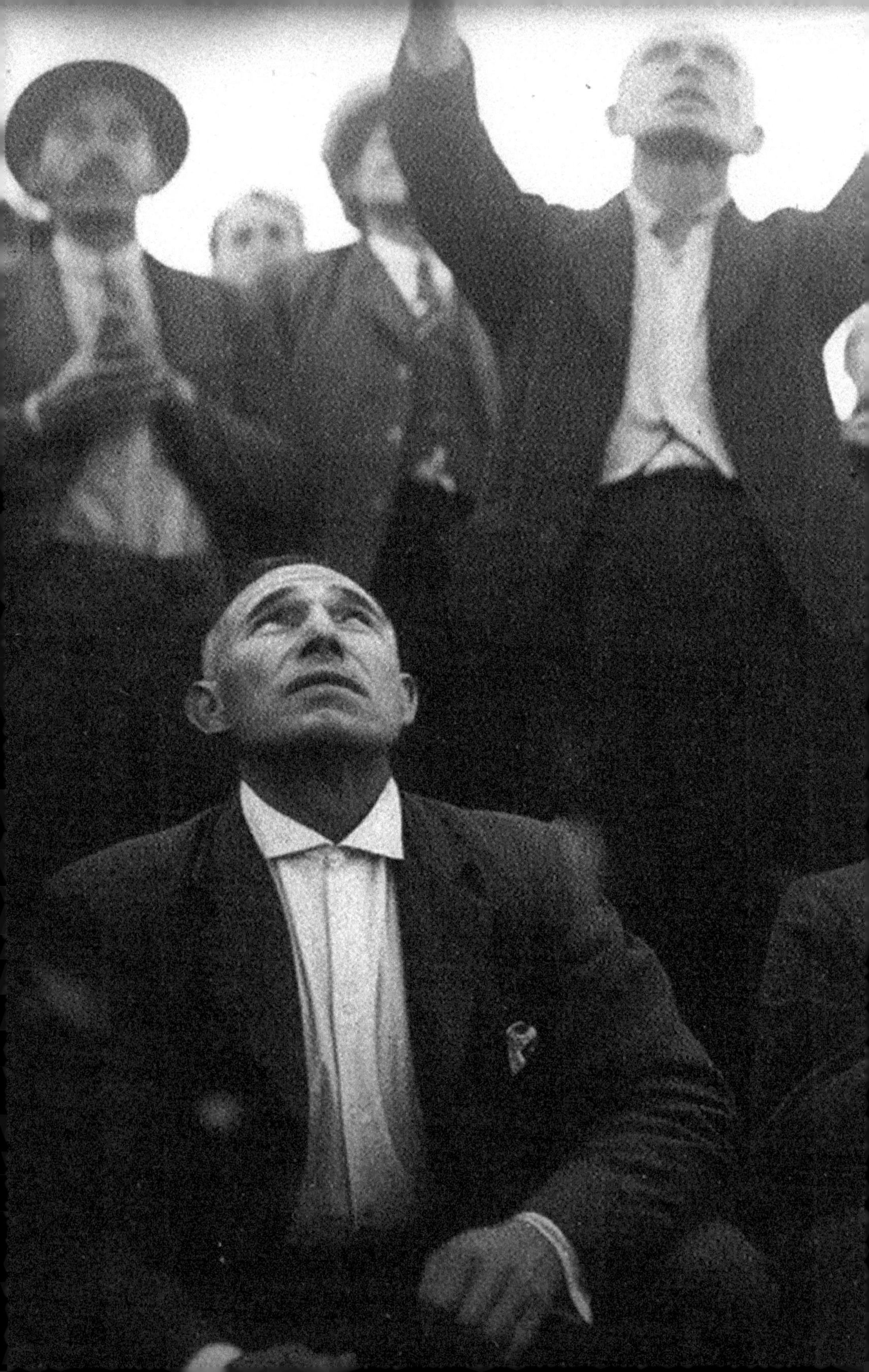

Impossible Images

AI as Iconography

By Mark Burchick

We are surrounded by images of the impossible. As Carlos Eire notes in "They Flew: A History of the Impossible," depictions of levitations, stigmata, and other miraculous phenomena not only reflected belief but also led to sympathetic phenomena being actualized by subsequent mystics.[1] These artworks were, and are, not innocuously consumed; seeing leads to believing leads to embodying. Even today, impossible events continue to be recorded, challenging our understanding of the natural world and generating beliefs.[2] They are not supposed to happen, *and yet they do*. It appears that images are somehow involved in this.

In 2022, a new kind of impossible image emerged. Programs such as DALL-E, Midjourney, and StableDiffusion dominated social media with artworks rendered through non-human intelligence. Filmmaker Bennet Miller's early experiments with DALL-E highlighted AI's unique ability to conjure photorealistic depictions of fictional events, framed by eyes and processed by hands that do not exist.[3]

These images carry spectral traces, or "hallucinations," of their training data. Artist and scholar Eryk Salvaggio describes AI art as a "haunted medium," whose compositions reveal both the data they derive from and the absences they inherit.[4] In this way, AI art becomes an "infographic for its dataset,"[5] unintentionally mapping the systemic

---

1. Eire, "They Flew. A History of the Impossible," 140.

2. See Kripal, *The Flip: Epiphanies of Mind and the Future of Knowledge,* and Pasulka, *Encounters: Experiences with Non-Human Intelligences.*

3. "Gagosian to Exhibit New AI-Generated Artworks by Bennett Miller in New York."

4. Salvaggio, "The Ghost Stays in the Picture, Part 1: Archives, Datasets, and Infrastructures."

5. Salvaggio, "How To Read an AI Image."

cultural and aesthetic decisions that shape our world and our perceptions.

Salvaggio, a member of the Algorithmic Resistance Research Group (ARRG!), calls for critical engagement of these tools, "connecting the critique and analysis of AI systems *to the use of AI systems.*"[6] I've taken a similar approach in my work, often analyzing AI's output through a religious and iconographic lens. As many popular media depictions will attest, when we speculate about the future of AI, we often label it with religious terminology, such as a savior to humankind or an apocalyptic forebearer.

I have written elsewhere about the potential for AI Art to be an interdisciplinary research tool, bridging art and the humanities in the deciphering of internet biases.[7] In developing this research, I was made to understand by my writing partner that my creative and research endeavors would become part of the material culture of the Catholic sacred tradition. It was not enough to merely analyze the output of AI systems. My work would now exist within and contribute to older traditions, religious and otherwise.

While I remain cautious of this technology, I acknowledge it as the logical result of a certain continuity in art history. In many art forms, we utilize technology to mediate how the "truth" is made known to our audience. By using AI to render impossible religious events, I am analyzing questions surrounding representation and authorship, and how these answers might structure our epistemologies.

With each passing year, the development of these technologies, government policy, and societal adoption are affirming that this new paradigm may be inevitable. And yet, we are still held in liminal tension. If images can *literally* make real the impossible, what can AI-generations tell us about our past, and what might they help us to imagine in the future?

---

6. Salvaggio, "The Algorithmic Resistance Research Group (ARRG!)"

7. Burchick and Pasulka, "Choreographing Shadows: Interdisciplinary Collaboration to Orchestrate Ethical AI Image-Making."

Images as Relics

Given longstanding traditions that artworks are believed to "capture" the soul, essence, or aura of their subject,[8] or in some cases, can assert an "aura" in and of itself,[9] I first considered the extent to which AI-generated images, and all images, function like relics.

Catholic relics, such as fragments of saints' bodies or belongings, are venerated for their *praesentia*, or a living "felt presence" of the deceased.[10] Many traditional practices that informed art history arose to venerate these relics, including pilgrimage to a shrine or burial site, housing them in highly decorated art objects called reliquaries, and the creation of portraiture or narrative images displayed alongside them.[11]

Most notably, as the tradition developed, the images themselves were believed to contain the presence of the saint regardless of presentation alongside relics, a tradition that is observed most prominently in Orthodox Christianity. However, this drew concern that the image would become the locus of worship, prompting the Byzantine iconoclasm and the destruction of such images. [12]

It is also for this reason that iconographic images became caught up in the age-old conundrum of representation as a distortion of reality. Some of Christianity's most important religious works change the details of their stories to promote aesthetic or theological emphasis. For example, Giotto depicts St. Francis of Assisi riding in a flaming chariot through the sky.[13] However, the initial primary text upon which this depiction is based mentions only an illuminated chariot passing through a house—neither Francis nor the night sky appears in the narrative.[14]

---

8. Strother, "'A Photograph Steals the Soul': The History of an Idea."

9. Benjamin, "The Work of Art in the Age of Mechanical Reproduction."

10. Brown, *The Cult of the Saints: Its Rise and Function in Latin Christianity*, 50.

11. Kreuger, "The Religion of Relics in Late Antiquity and Byzantium," 8-9.

12. Ibid, 10-12.

13. "Legend of St Francis: Vision of the Flaming Chariot."

14. Qtd in Armstrong, "Francis of Assisi: Early Documents - The Saint," 224.

This derivation reflects Catholic theological developments that cast Francis as a prophet, aligned with earlier biblical visions. [15]

These hagiographic depictions are meant, in part, as educational tools for an illiterate audience, aestheticizing the story with emotionally resonant and archetypal imagery and drawing on the historical artworks of the tradition. They communicate their message through simplified and symbolic language.

In the contemporary era, Andy Warhol's pop art channels the continuity of this "felt presence." Raised Catholic, Warhol absorbed the hallmarks of Byzantine iconography. His silkscreen portraits appropriated pre-existing images of Marilyn Monroe and others, replicating the style of devotional images and elevating secular figures into pop icons. [16] Despite the mass production of these works, they generate and retain significant value because of Warhol's cult-like presence and influence in the contemporary art world. They exude the "felt presence" of the artist. The image is not just of Marilyn Monroe, but *it is a Warhol.*

This trajectory highlights Salvaggio's observation that AI-generated images are shaped by systemic structures. Programs like Midjourney and DALL-E are deliberately tweaked to favor broadly commercial and visually appealing imagery within their training sets, a process called "aesthetic weighting."[17] Like theologians and iconographers before them, AI engineers influence what can be imagined and what is excluded.

An image prompt for "Jesus," for example, will almost always produce a white, European-appearing man, reflecting long-standing art traditions that obscured the historical Jesus's Middle-Eastern origins. A single AI image encapsulates how a concept is popularly understood, a "truth" created through centuries of decisions by artists, patrons, curators, editors, developers, and engineers who have deemed certain images permissible and others not.

---

15. Armstrong, "Francis of Assisi: Early Documents – The Prophet," 11.

16. Kochetkova, Marina. "Andy Warhol: Religious Artist for a Secular Society."

17. Salvaggio, "Critical Topics: AI Images, Class Nine: Exploring the Datasets."

The Unseen "Hand" of AI

Even more striking is the question of AI-image authorship. When a technology produces an image, where does it come from? Chesher and Albarran-Torres note that the black-box nature of popular AI image technologies produces an effect whereby the instantaneous production of an image from text suggests a magical invocation, working with a non-human intelligence as one might channel in a séance or through a Ouija board. This is confirmed by users who describe the technology as spooky, eerie, uncanny, and awe-inspiring.[18] The AI-generated image becomes a vessel or mediator to a perceived non-human intelligence.

The images that are produced, no matter how photorealistic they may appear, are often made glamorous, with careful attention paid to composition and lighting. Frequently, though, it is the perspective of the impossible author that betrays their facade. Unnatural camera angles, uncanny spatial relationships, and any manner of invented phenomenon depicted defy our expected cultural visual literacy.[19] These subversions highlight the question of the image's authorship.

We can account for the training set and code that produces an image, but we can't quite explain the "source" of an AI's "creativity," which is a qualifier we apply to the image during the analysis. Whether the AI or its human-prompting counterpart is credited as the author of the image, this contentious relationship provides space for speculation about their motives and origins.

Within religious art history, we have a category for such images. *Acheiropoieta* is a Greek term which means "made without human hand," and these images are an important part of the iconographic tradition. Within Catholicism, some notable examples include the Shroud of Turin, which is believed to be Jesus's burial cloth and shows a representation of an entombed body, as well as the Our Lady of Guadalupe image, a portrait of the Virgin Mary said to have been miraculously emblazoned on a peasant's cloak in the 1530s.

---

18. Chesher and Albaran-Torres, "The Emergence of Autolography: the 'Magical' Invocation of Images from Text through AI," 3-4.

19. Ibid, 7.

These images are claimed by their proponents to be the result of spontaneous non-human processes, a kind of divine light that instantly recorded the visage of Jesus or the Virgin Mary on light-sensitive organic fabrics.[20] These artifacts are controversial because both science and faith have weighed in to explain their origins, resulting in a tug-of-war negotiating how these images are understood among religious and secular people alike, and how they reinforce belief systems.

AI-generated images invite similar scrutiny, and rightfully so. Much concern has been paid to issues revolving around data-scraping and privacy practices, copyright infringement, harms posed by their photorealistic qualities, and the absence of a "human touch" that would elevate the work to the realm of art.

While I sympathize with these concerns, some of them mirror historical reactions to past technologies, especially against the introduction of the handheld camera in the nineteenth century.

One anonymous critic, writing in the same year the daguerreotype was released to the public, said of the technology that "All nature, animate and inanimate, shall be henceforth its own painter, engraver, printer and publisher… and yet, how unsatisfied people are with it, because it is not of human hand."[21] The implication is that because the technology was believed to reproduce nature, there was no actual artistry being done and no capability of deception, discounting the "hand of the artist," but ignoring the compositional eye mediated by technology.

Although a prompt can generate an AI image in mere seconds, the pressing of the "Enter" key conceals decades of cultural and computational decisions in the act. Like acheiropoieta, these images arise from nebulous authorship and then are defined by how they are contextualized, blurring the line of what is considered "human-made." Are they revelation or manipulation?

---

20. See Sennott, *Acheiropoeta: Not Made by Hands: Miraculous Image of Our Lady of Guadalupe and the Holy Shroud of Turin.*

21. Parker Willis, Nathaniel, and Timothy O Porter. "The Pencil of Nature. A New Discovery," 45-47.

Fatima

The body of work that I have produced in response to this research, "Felt Presence," explores the 1917 Marian Apparitions in Fatima, Portugal. Three children, Lucia Dos Santos and her cousins Francisco and Jacinta Marto, claimed multiple visitations by the Virgin Mary and angelic figures. Mary is alleged to have given the children prophetic visions and promised a miracle would take place on October 13th of that year. Local journalists reported the story, and crowds at these events grew extensively, until tens of thousands were present at the culminating apparition.[22]

Eyewitness testimony of these events, including that of secular and skeptical sources, such as lawyers, doctors, and journalists, is voluminous, and they record two primary miracles: flower-like objects which fell from the sky[23] and the sun "dancing" at the October 13th apparition.[24]

Judah Bento Ruah, a photojournalist working for the weekly secular journal *Illustração Portuguesa,* documented the crowds in 13 glass negative plates, but captured neither of these miraculous events.[25] According to an analysis of the photos taken in 2021, Bento Ruah would have likely not been prepared for photographing the skies, as the technical limitations of the era would have required specially prepared equipment, including lenses, filters, and treated plates.[26]

The newspaper where the images were first presented was only made possible by the advent of new technologies at the turn of the century, namely easy-to-use, portable Kodak cameras and photoengraving, a method of industrial photo printing. Considering that Portugal had an illiteracy rate of almost 70% in 1917, Ruah's photographs were informed by the journal's editorial ethos, constructing a narrative through

---

22. Łaszewsk, *The World of Marian Apparitions,* 28 and 34.

23. Fernandes and D'Armada, *Heavenly Lights: The Apparitions of Fatima and the UFO Phenomenon,* 84.

24. Arcement, "Our Lady of Fatima: The Virgin Mary promised three kids a miracle that 70,000 gathered to see."

25. André Melícias, e-mail correspondence, October 11, 2023.

26. Catrica, "As 13 fotographias de Judah Bento Ruah do dia 13 de outubro de 1917," 22.

sequential images meant to be seen first, and then read afterwards.[27] Like the medieval icons that came before, researcher Paulo Catrica observes that the photographs are "inseparable from the historiography of the Apparitions of Fatima," establishing a "multiplicity of symbolic meanings," which evolve with time.[28]

The disconnect between overwhelming testimony and the absence of physical evidence embodies the dilemma of how best to represent these impossible events. It is my personal belief that these events took place as described, not because I dismiss the scientific and societal pressures that seek to negate the reality of such events, but because this archive has played a surprising geopolitical role in the history of the Twentieth and Twenty-First centuries.[29] Faith and reason support their legitimacy even in the absence of proof.

The Felt Presence of AI Images

By feeding Ruah's images into Midjourney, the program most designed to alter a composition through "aesthetic weighting" parameters, I produced "imagined realities" of the Fatima miracles, rendering photorealistic, period-appropriate images of the "Miracle of the Sun" and flower petal shower.

I then transferred these digital image assets, both Ruah's original images and my AI-generated ones, onto glass plate negatives, mirroring the original format in which they were photographed and harkening their "relic-like" status, having memorialized the miraculous. These backlit negatives, alongside positive silver-gelatin prints, are presented together, with no notification that some are AI-generated. They are all attributed to Judah Bento Ruah, a photojournalist on assignment from a Portuguese newspaper.

Displayed in museum curatorial fashion, using historical photographic processes, the aesthetics of credibility are pushed to a breaking point. By analyzing *events that cannot happen*, the viewer must

27. Ibid, 150-163.

28. Ibid, 163.

29. See Jan Margry, *Cold War Mary: Ideologies, Politics, and Marian Devotional Culture.*

engage the images critically. Is what I am seeing true, and what does "truth" even mean in our fraught era, when photorealistic historical imagery can be generated by contemporary technologies.

AI data makes it nearly impossible to differentiate between real and fake information and human versus machine-generated interactions.[30] The very existence of these technologies upends every notion we ever had about faith in the art form to express that which is grounded in truth.

My deception is intentional but not malicious. The project exists to highlight a unique tension: the impossible does happen, and we must mediate that knowledge through fallible humans and technologies. As an artist, researcher, and person of faith, I cannot ignore the fact that there are control systems that establish how we understand our reality, and they all fall short of what they are designed to do.

I have a responsibility to educate and inform, but the use of AI imagery in this work comes with an ethical burden; namely, do I reveal the "magician's trick" or do I make the audience work to discover the illusion themselves? I want the viewer to experience the unease of uncertainty

Susan Sontag presaged our current media landscape by comparing it to Plato's allegory of the cave, stating that "the picture may distort, but there is always a presumption that something exists, or did exist, which is like what's in the picture."[31] She goes so far as to call us "image junkies… needing to have reality confirmed and experience enhanced by photographs."[32] In a post-AI era, we are threatened by an image overdose.

Mirroring Catrica's analysis that the Fatima images have changed their meaning over time, I determined that my work would challenge an audience's ability to discern, and therefore, would provide an invaluable tool they will need to navigate our landscape of post-AI images. We must become, as the art critic Clement Chéroux puts it, practitioners of a "new hygiene of vision."[33]

---

30. Walter, " Artificial influencers and the dead internet theory," 1.

31. Sontag, "On Photography," 5.

32. Ibid, 24.

33. Qtd in Fontcuberta."Camouflages," 26.

Scarce wall text offers witness testimony to contextualize what is seen in the images and to develop a chronological narrative. But for those unfamiliar with the Fatima event, the images may appear to be an invented story, albeit a convincing one.

Any research into the topic will validate what is seen. The viewer is caught in a paradox: How does one make sense of images *that intellectual reasoning tells them must be impossible*? They must lean into discernment to determine what is real and what is fake, and why someone might seek to engineer their beliefs in such a way.

What our history of impossible images reveals is that there is a critical relationship between cultural depictions and individual experience. AI art highlights this relationship by forcing us to reconcile how images, mediated by technology and power structures, build aesthetic traditions, belief systems, and our perception of reality. In doing so, we reflect on how we arrived here and determine how best to proceed.

Who are the people behind the images who seek to capture and control our attention? How are images used to affect our beliefs? These are not rhetorical questions, but invitations critical to our times. If we are to navigate a treacherous sea of AI images, we must sharpen our discernment, not just to determine what is true, but to ask what might be possible.

References

Arcement, Katherine. "Our Lady of Fatima: The Virgin Mary promised three kids a miracle that

70,000 gathered to see." *Washington Post,* October 13, 2017. https://www.washingtonpost.com/news/retropolis/wp/2017/10/13/our-lady-of-fatima-the-virgin-mary-promised-three-kids-a-miracle-that-70000-gathered-to-see/

Armstrong, Regis J. "Francis of Assisi: Early Documents - The Saint." Franciscantradition.org. Accessed July 17, 2025. https://digitalcollections.franciscantradition.org/document/bx4700-f6f722-1999/francis_of_assisi_early_documents_-_the_saint/1999-00-00?pageNo=224.

Armstrong, Regis J. "Francis of Assisi: Early Documents – The Prophet." Franciscantradition.org. Accessed July 17, 2025.https://digitalcollections.franciscantradition.org/document/bx4700-f6f722-1999-v3/francis_of_assisi_early_documents_the_prophet/1999-00-00?pageNo=11

Benjamin, Walter. "The Work of Art in the Age of Mechanical Reproduction." In *Illuminations,* edited by Hannah Arendt. Schocken Books, 1969.

Brown, Peter. *The Cult of the Saints: Its Rise and Function in Latin Christianity.* The University of Chicago Press, 1981.

Burchick, Mark and Pasulka, Diana Walsh. "Choreographing Shadows: Interdisciplinary Collaboration to Orchestrate Ethical AI Image-Making," *Tradition Innovations in Arts, Design, and Media Higher Education* 1, no. 1 (October 15, 2023). Accessed at: https://digitalscholarship.unlv.edu/tradition_innovations/vol1/iss1/2

Catrica, Paulo. "As 13 fotographias de Judah Bento Ruah do dia 13 de outubro de 1917." In *Fatima XXI,* 8, (November 2017): 148-165.

Chesher, Chris, and César Albarrán-Torres. "The Emergence of Autolography: the 'Magical' Invocation of Images from Text through AI." *Media International Australia,* 189, No. 1 (2023). doi:10.1177/1329878X231193252. Accessed at: https://www.researchgate.net/publication/373246708_The_emergence_of_autolography_the_%27magical%27_invocation_of_images_from_text_through_AI

Eire, Carlos M. N. *They Flew: A History of the Impossible.* Yale University Press, 2025.

Fernandes, Joaquim and Fina D'Armada. *Heavenly Lights: The Apparitions of Fatima and the*
UFO Phenomenon. Translated and Edited by Andrew D. Basiago and Eva M. Thompson. Anomalist Books, 2007.

Fontcuberta, Joan. *Camouflages.* Contrasto, 2014.

"Gagosian to Exhibit New AI-Generated Artworks by Bennett Miller in New York." *Gagosian*, March 7, 2023. https://gagosian.com/media/exhibitions/2023/bennett-miller/Bennett_Miller_2023_Gagosian_976_Madison_Avenue_New_York.pdf.

Jan Margry, Peter. *Cold War Mary: Ideologies, Politics, and Marian Devotional Culture.*
Leuven University Press, 2021.

Kreuger, Derek. "The Religion of Relics in Late Antiquity and Byzantium." In *Treasures of*
Heaven: Saints, Relics, and Devotion in Medieval Europe. Edited by Martina Bagnoli, Holger A. Klein, C. Griffith Mann and James Robinson. Yale University Press, 2010.

Kripal, Jeffrey. *The Flip: Epiphanies of Mind and the Future of Knowledge.* Bellevue Literary
Press, March 12, 2019.

Kochetkova, Marina. "Andy Warhol: Religious Artist for a Secular Society." DailyArt Magazine, August 6, 2024. https://www.dailyartmagazine.com/andy-warhol-religious-artist/.

Łaszewski, Wincenty. *The World of Marian Apparitions.* Sophia Institute Press, September 21,
2021.

"Legend of St Francis: Vision of the Flaming Chariot." *Web Gallery of Art.* Accessed July 17, 2025. https://www.wga.hu/html_m/g/giotto/assisi/upper/legend/franc08.html.

Parker Willis, Nathaniel, and Timothy O Porter. "The Pencil of Nature. A New Discovery." In *Photographic Theory: An Historical Anthology*. Edited by Andrew E. Hershberger. Wiley Blackwell, 2014.

Pasulka, Diana Walsh. *Encounters: Experiences with Non-Human Intelligences.* St. Martin's Essentials, November 7, 2023.

Salvaggio, Eryk. "How To Read An AI Image." *Cybernetic Forests,* Oct 2nd, 2022. https://cyberneticforests.substack.com/p/how-to-read-an-ai-image

Salvaggio, Eryk. "Critical Topics: AI Images, Class Nine: Exploring the Datasets." *Cybernetic Forests,* May 1, 2023. https://www.cyberneticforests.com/ai-images-9

Salvaggio, Eryk. "The Algorithmic Resistance Research Group (ARRG!)" Cybernetic Forests, Aug 13, 2023. https://cyberneticforests.substack.com/p/the-algorithmic-resistance-research

Salvaggio, Eryk. "The Ghost Stays in the Picture, Part 1: Archives, Datasets, and Infrastructures." *Flickr,* 2024. https://www.flickr.org/the-ghost-stays-in-the-picture-part-1-archives-datasets-and-infrastructures/

Sennot, Thomas Mary. *Acheiropoeta: Not Made by Hands: Miraculous Image of Our Lady of Guadalupe and the Holy Shroud of Turin.* Ignatius Press, 1999.

Sontag, Susan. *On Photography.* Picador, 1973.

Strother, Z.S. "'A Photograph Steals the Soul': The History of an Idea." In *Portraiture and Photography in Africa.* Edited by John Peffer and Elisabeth Cameron. Indiana University Press, 2013.

Walter, Yoshija. "Artificial influencers and the dead internet theory." *AI & Society* 39, No. 1 (February 5, 2024). Accessed at: https://link.springer.com/article/10.1007/s00146-023-01857-0

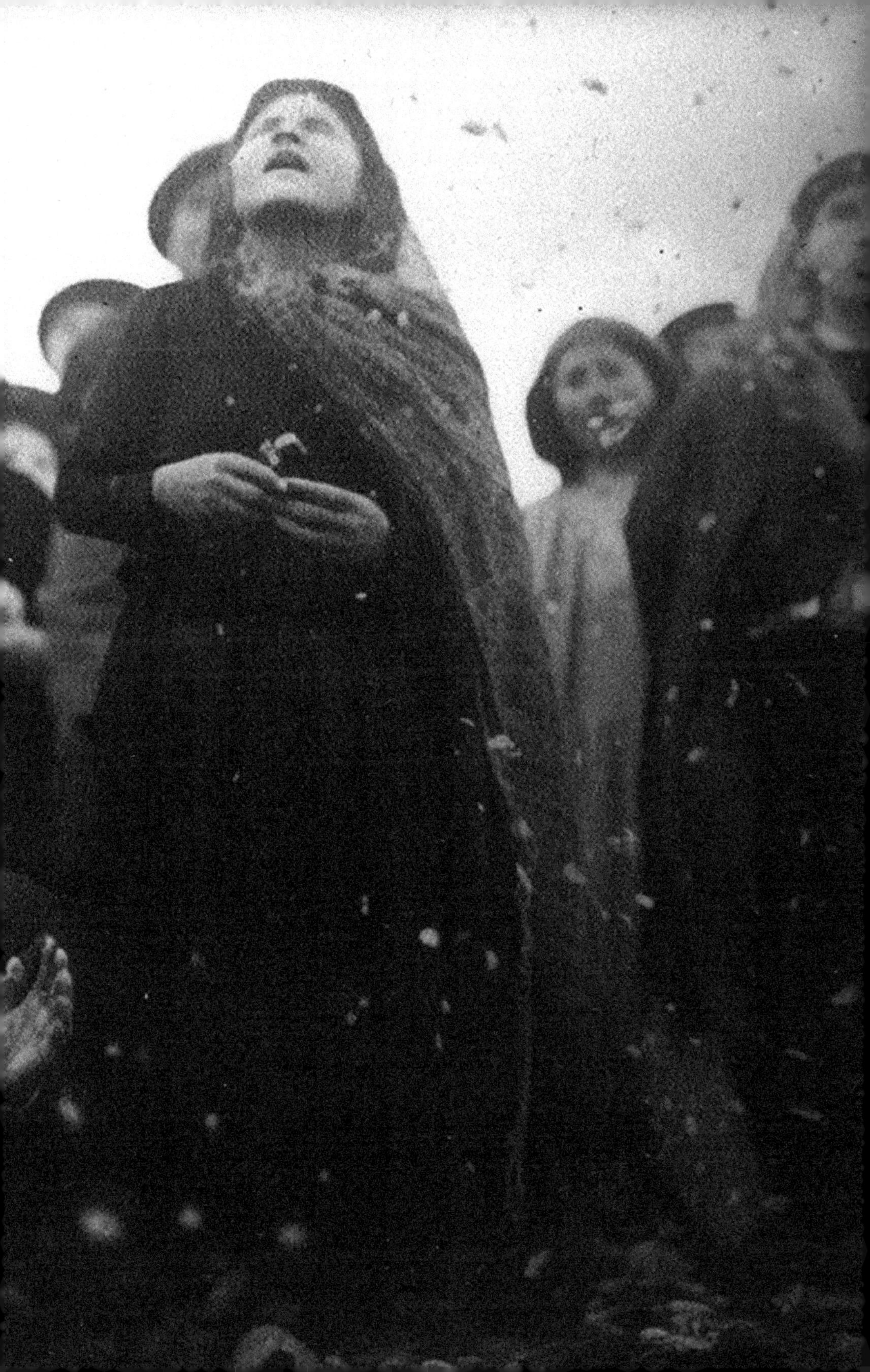

Emergent Ecologies of AI-Driven Art

Rethinking Authorship, Agency, and Process

By Bei Hu

We are living in a pivotal moment in history, where the speed of AI-powered platforms is unprecedented. It's like riding a roller coaster, where everyone feels the rush of wind as it speeds through the air. For creative practitioners, our workflows have been altered on various levels, both positively and negatively. Please buckle up, as we anticipate that change will continue to come rapidly in the months and years ahead.

In this essay, I aim to investigate how AI has impacted the creative industry and how artists and designers are utilizing it, exploring their workflows and motivations. By examining existing examples of work and effective patterns of artistic production with AI, I hope to assist fellow creative practitioners in seeing the diverse paths available for their own AI-driven creative production. For those who don't necessarily need to learn about AI for their creative work, this paper may help you articulate the uniqueness of your creations, especially in a context where your audiences have questions about authorship, originality, and the value of artwork.

Like all technologies and tools from the past, we must acknowledge that using them for our creative production may not automatically add value to our work but may attract temporary attention from audiences, which should not be our goal. Additionally, I aim to understand the factors that influence the value of artwork and our sense of agency, particularly concerning the value that AI may add to our work.

The term 'ecologies' I used in this essay refers to the interconnected systems that encompass not just the artworks themselves, but also the relationships between artists, audiences, artworks, and machines. It suggests that the creation of art in the context of artificial intelligence is a dynamic and interdependent network. This ecosystem is influenced by various factors, such as data, human interactions, and adaptive algorithms, which all play a role in shaping the artistic experience.

In later sections of this essay, I discuss the boundaries between human and machine creativity, and point out the missing links and dimensions within this new ecosystem that AI fails to comprehend.

This structure of the essay follows the 3 research questions that I ask:

R1 How has AI been integrated into contemporary art (and design) practices, and what significant shifts have occurred in the past year?
R2 Within these AI-driven art practices, what new ecologies have emerged, and what ethical considerations have they raised?
R3 How does collaborative artmaking with AI redefine the boundaries between human and machine creativity, and how can we identify aspects of creative practice uniquely suited to human ingenuity versus those effectively delegated to AI systems or collaborators?

R1: 1. Current AI-powered platforms have lowered barriers to creating visual, audio, and textual content:

In early 2025, from February to April, the number of weekly active users of ChatGPT doubled from 400 million to 800 million within just two months.[1] This surge marked a significant moment for AI, as it entered the mainstream through chatbot interfaces.

Recent advancements in generative AI have reduced the technical barriers and expertise needed to create high-quality, AI-powered content. For instance, in the first week following the launch of ChatGPT's upgraded image generator on March 25, 2025, 130 million users created over 700 million AI-generated images.[2] This represents a remarkable leap compared to the 1.5 million early adopters of DALL·E, who were generating about 2 million images per day by September 2022.[3]

Reflecting what you and I experienced in early 2025, individuals and families around the world now create image and video content directly on their phones to share with loved ones, communities, and

1. Singh, *ChatGPT Statistics & Total Users (2025): DAU & MAU Data.*

2. Wiggers, *ChatGPT Users Have Generated over 700M Images since Last Week, OpenAI Says.*

3. *DALL·E Now Available without Waitlist.*

broader audiences through social media. These are not automatic artworks, but somehow they spark people's minds to imagine a static image as a video, and transform a photo into the illustration style of Hayao Miyazaki.

I received a video clip of my father created from one of his old, faded black-and-white photographs. It depicted my parents on their wedding day. In the short video, the youthful couple turned toward each other and smiled at the camera.

Unlike many other videos he had made, which often resulted in strange outputs such as blurred hands or distorted faces at certain angles, this video of my parents was quite successful; it appeared realistic and vivid, transporting us back to that moment and allowing us to experience life through video clips instead of just still images. This recreation of everyday moments highlights a significant turning point: AI tools have become more accessible to the general public for specific use in their everyday lives. Through conversational interfaces on websites or apps, anyone with an internet connection can now participate in AI-powered content production.

R1: 2. AI augments and accelerates production workflows in the Art and Design Industry

In recent years, the integration of artificial intelligence (AI) has transformed how artists and designers approach their work. A prominent question that has arisen is: What will the labor economics of creative work look like in the future? Specifically, which aspects of creative work will be replaced by AI, what new positions will be created, and how will this shift affect the general public's perception of the value of certain creative productions?

AI is being used across various creative fields, such as graphic design, fashion, architecture, and multimedia art. It helps streamline processes, optimize production workflows, generate ideas, and create visual, audio, and text content. Additionally, AI aids in knowledge retrieval and decision-making, allowing creatives to concentrate more on innovative thinking and higher-level concepts. By leveraging AI's capabilities, they can enhance efficiency and improve the quality of specific workflows.

According to the "AI in Content Creation 2025" report, nearly 40 percent of creators now use AI throughout their entire workflow, from initial drafts to final production and distribution.[4] The Adobe Creative Frontier Survey indicates that 90 percent of creators believe generative AI tools help relieve them of menial tasks, ultimately saving time and costs. However, this report also reveals a mixed perception of AI, with people expressing both optimism and concern. Concerns include the potential misuse of AI—specifically, the possibility that individuals' work could be exploited to train AI models without their consent. In response to these concerns, 88 percent of respondents support new laws to protect creators' work from imitation by AI tools.[5]

As AI continues to evolve, the balance between leveraging AI and maintaining artistic integrity remains a topic of significant discussion within the creative community.

R1: 3. The use of AI in the creative industry highlights the growing distinction between tasks that can be handled more efficiently by AI and those that are better suited for human expertise.

The use of AI in the creative industry underscores the increasing distinction between tasks that can be handled more efficiently by AI and those that are better suited for human expertise. Researchers suggest that AI should be viewed as a partner that assists human artists in exploring new artistic possibilities by providing alternative perspectives [6]or tackling challenges that are too difficult for humans, such as analyzing vast databases to draw conclusions. [7]At the same time, AI can relieve artists from low-judgment, time-consuming tasks that require minimal critical thinking, creativity, or subjective decision-making. This, in turn, allows

4. Wondercraft, *AI in Content Creation 2025, a New Report from Wondercraft, Reveals That over 80% of Creators Are Now Using AI in Their Workflows—Nearly 40% from Start to Finish.*

5. Adobe Communications Team, *Adobe's AI and the Creative Frontier Study Reveals Creators' Views on the Opportunities and Risks of Generative AI.*

6. Deckker and Sumanasekara, "A Review of AI-Powered Creativity: The Intersection of AI and the Arts," 18; Nicholas Davis and Janet Rafner, *AI Drawing Partner: Co-Creative Drawing Agent and Research Platform to Model Co-Creation.*

7. Anantrasirichai and Bull, "Artificial Intelligence in the Creative Industries: A Review," 636.

artists to concentrate on more complex and nuanced creative decisions. Both perspectives are valid. By leveraging AI in both high-level and low-level tasks, we can save time and resources across different contexts and objectives.

For designers and artists, it is crucial to understand how AI-powered systems are replacing traditionally human roles. This understanding allows us to identify where we should focus our energy, time, and passion for learning, while also being mindful of tasks that can be effectively delegated to AI.

However, it's important to remember that while AI can excel in certain aspects of art creation, it is ultimately up to the artist to decide how to integrate AI into their work and to what extent. The value of time-consuming, human-made tasks lies in their ability to create meaningful connections within their context. AI might not add any value at all to creative work if not positioned in a specific context for a purposeful goal.

For creative practitioners who may not necessarily need to work with AI, observing how AI is being incorporated into other artists' work can serve as a mirror. This reflection allows us to examine what remains valuable in our creativity as human beings. At the core of this inquiry is the question of what we can learn about our unique methods, thoughts, and creativity as individual creators. One of the most important values each creator can bring into the world is their uniqueness, combined with skills, knowledge, ethics, and a passionate heart.

R2 Within these AI-driven art practices, what new ecologies have emerged, and what ethical considerations have they raised?

Within these AI-driven art practices, what new ecologies have emerged? "Ecologies" refers to the new systems, relationships, and environments that have emerged with the advancement of AI-driven art practices. This includes not only the artistic creations themselves but also the interactions between AI and human artists, as well as the relationships among artists, audiences, and machines. These new dynamics involve how art is created, distributed, and consumed, along with the collaborative relationships between human artists and AI tools.

R2: 1. What's the dynamic interplay between human creativity and artificial intelligence in artistic endeavors look like?

To explore the collaborative relationship between human artists and AI in the creative process, we can use the analogy of an artist and their studio assistants. The assistant works closely with the artist and is highly skilled at specific tasks, but it does not possess agency, emotions, or self-awareness.

1. Project Management

The artist uses their intellect to break down problems into smaller, manageable steps, then delegates specific tasks to the AI and ultimately assembles and synthesizes these components in a way that aligns with the overall project. The assistant faithfully handles all assigned tasks, while it is the artist who determines what work to delegate—sometimes high-level tasks and other times time-consuming, low-level work. The key difference here is that the artist assigns tasks to the assistant rather than receiving tasks from it. The artist is also responsible for knowing which types of work to assign to different AI assistants, and if multiple assistants are involved, the artist facilitates communication between them.

2. Art direction

While the assistant may offer suggestions for artistic directions, the artist ultimately directs the assistant's efforts, relying on their training, intuition, unique taste, and combined knowledge. The artist makes crucial decisions while remaining open-minded and receptive to feedback.

Ultimately, the artist has the final say on whether to accept or reject the assistant's suggestions. Although AI can surprise us with its capabilities, we must be cautious, as the assistant, despite its best efforts, can still make mistakes. Therefore, quality control is an essential part of the artist's role.

3. Vision

The artist casts a vision and provides leadership, maintaining a broader perspective while understanding the deep purpose and value driving their work. They grasp the rationale behind each decision and envision larger goals worthy of the time and energy invested. Artists can see the entire forest rather than just a jungle, enabling them to navigate effectively toward their objectives. In many cases, their work carries a significant purpose, helping them respond to specific subjects or issues in the world through their creative practice.

There are various touchpoints throughout the creation process, from exchanging initial ideas to refining specific workflows for final production. In this partnership, humans take the lead, with AI serving as a supportive teammate that follows the artist's direction and vision.

The ecosystems surrounding AI-driven art production and its audiences raise important questions that challenge our current paradigms. Specifically, how does AI impact the value of an artwork? What metrics can we use to define whether the value has increased or decreased due to the incorporation of AI models? How can we ensure that the general public perceives this art as valuable? Ultimately, the conversation centers on issues of transparency and trust.

It also brings up new discourse to questions that has long been discussed in art, such as originality and authorship, although every new medium in art history rekindles debates about originality and challenge paradigms, such as when first camera, "daguerreotype" first introduced, critics wondered if the photographer or machine deserved credit for the "art" being captured. For early computer art, Harold Cohen's AARON program generated drawings from code he wrote, which sparked debate about who is the real artist, the programmer or the program? While AI revives familiar debates about originality and authorship, but on a much bigger, less transparent scale.

If we compare an AI model with programs in early computer Arts. The AI model is more opaque than transparent to the artists and designers working with it. For example, for the case of AARON, Cohen is the ultimate architect of his artistic process; he designed the program to serve him as an extension of his hand and mind, the program is transparent to him, and entirely under his control. In contrast, most artists and designers can't fully understand the AI model behind the interface they have been interacting with. AI Models are trained on massive, unseen datasets; the "reasoning" is opaque.

AI introduces new dimensions of creative interaction. Building on a lineage of structured, rule-based approaches, from John Cage's chance operations and Sol LeWitt's conceptual instructions to generative art's algorithmic methods, AI-driven art differs significantly because it uses learning models that evolve and adapt over time. This can be observed in several ways:

1. Learning Process: AI learns differently than humans.
2. Audience Interaction: AI interacts with audiences in unique ways.
3. Production Methods: AI produces art differently.

AI has a large, data-driven foundation to learn from, drawing on diverse sources. This allows it to create unexpected combinations that incorporate a wide range of styles, influences, and techniques. Additionally, AI can offer a more personalized experience for users by analyzing preferences and feedback in real time. Moreover, the scalability of AI-driven platforms is remarkable. They can generate numerous variations of artifacts or specific content based on a single prompt, far exceeding the output of an individual artist or the productivity of a studio or creative agency.

Historically, rule-based generative art practices focused on carefully designed conditions that produced emergent and often unpredictable outcomes, ultimately redefining the relationship between process and product. By integrating adaptive machine learning models, AI-driven art transcends static rules, evolving into a living entity that changes based on human presence, viewer input, and dynamic environmental factors.

In the interactive installations titled "Appropriate Response (figure 1)," [8]AI art pioneer Mario Klingemann employs the GPT-2 neural network to generate texts that resemble aphorisms. These texts are showcased on a vintage split-flap display. The model was trained on 60,000 quotes sourced from the internet, and each new quote is generated when audience members kneel on a custom-built kneeler in front of the installation.[9]

The artist's practice is grounded in treating AI not merely as a tool but as a creative collaborator. He utilizes AI for text-based content generation to such an extent that audiences cannot discern whether a quote was written by a machine or a human, challenging their perceptions. His work highlights the blurred lines between human-created and machine-generated content, setting the stage for his "AI performer" to interact with the audience.

---

8. Mario Klingemann, *Appropriate Response*, August 29, 2020.

9. Mario Klingemann, *Appropriate Response by Mario Klingemann*; Mario Klingemann, *'Appropriate Response,'* 2020.

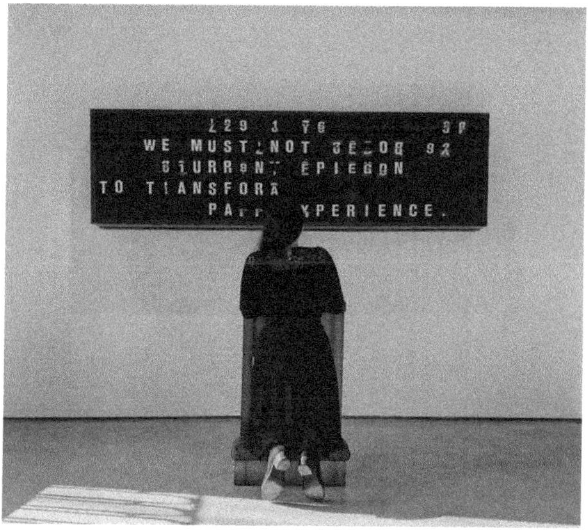

Figure 1. *Appropriate Response,* Mario Klingemann

Today's AI ecosystems provide adaptable "translation" pipelines between different media. Many artists utilize text-to-image, or text-to-video generation, which often surprises people when the typical approach is reversed. One notable example is the Poetry Camera (figure 2)[10], an art project developed by Kelin Carolyn Zhang and Ryan Mather. This camera analyzes what people capture and then prints a short poem based on what it "sees."

In what ways should creators declare their use of AI to build trust with their audiences while also protecting their unique ideas about how AI could be implemented? One effective way to foster this trust is by clearly explaining the intention and motivation behind using AI as a collaborator, detailing why it is being utilized and how it makes a difference.

In his recent exhibition at Thaddaeus Ropac in London, American artist David Salle unveiled "Some Versions of Pastoral", [11]a collection of paintings created with the assistance of AI. He overlays his hand-painted brushstrokes onto the AI-generated variations based on his previous work.

In an interview, he emphasized the vital role of the artist in "inhabiting the ambiguity" of originality.[12] Salle's art demonstrates how collaborating with AI allows him to explore compelling questions, fostering trust and interest among his audience. By posing questions that intrigue both himself and his viewers, his work transforms into an experiment and a laboratory for exploring new forms of creative collaboration. This artwork blurs the lines between human and machine

---

10. Kelin Carolyn Zhang and Ryan Mather, *Poetry Camera.*

11. David Salle, *Some Versions of Pastoral.*

12. *Thaddaeus Ropac Exhibition: Some Versions of Pastoral.*

authorship through a hybrid analog-digital workflow with hand-painted interventions over the AI-generated base image.

When audiences understand the reasons behind and the methods used by artists who incorporate AI, it helps to contextualize their creations. This understanding allows audiences to appreciate the unique value that an AI-human collaboration brings to the art-making process, distinguishing it from traditional collaborations with studio assistants or human partners.

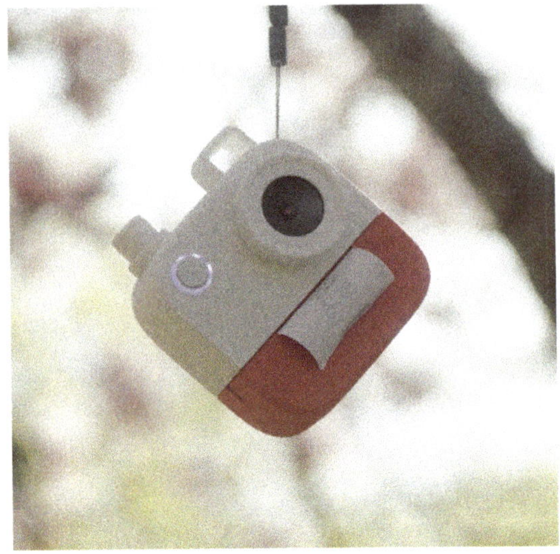

Figure 2. *Poetry Camera,* Kelin Carolyn Zhang & Ryan Mather

In recent conversations with artists and designers using AI, I gained insights into their diverse perspectives on the relationship between their creative practices and technology, along with their distinct workflows. Professor Tiffany Calvert, MFA chair in Visual Art at the Sam Fox School of Art, connects the history of painting to our current visual culture through her work. Her recent work includes a series of paintings (such as 464, figure 3[13]) layered over printed images generated by a custom machine-learning model trained on a dataset of more than 600 historical still-life paintings. She creates a dialogue between human-made abstraction and that generated by AI, where those AI-generated visuals serve as a foundation for her artistic expression.

Creative Director Étienne Mineur shared his fascinating workflow, starting with traditional methods—pen and pencil—before exploring AI models for various design options. This approach enables him to visualize different artistic directions and refine specific elements. For his Flore-AI (figure 4)[14], he integrated manual drawings into Midjourney, resulting in a dynamic 3D typeface with patterns transitioning from 2D to 3D.

Artist Sofia Crespo values AI's ability to extract patterns from

13. Calvert, *464*.

14. Mineur Etienne, *FlORA-AI*.

Figure 3. *464*, Tiffany Calvert

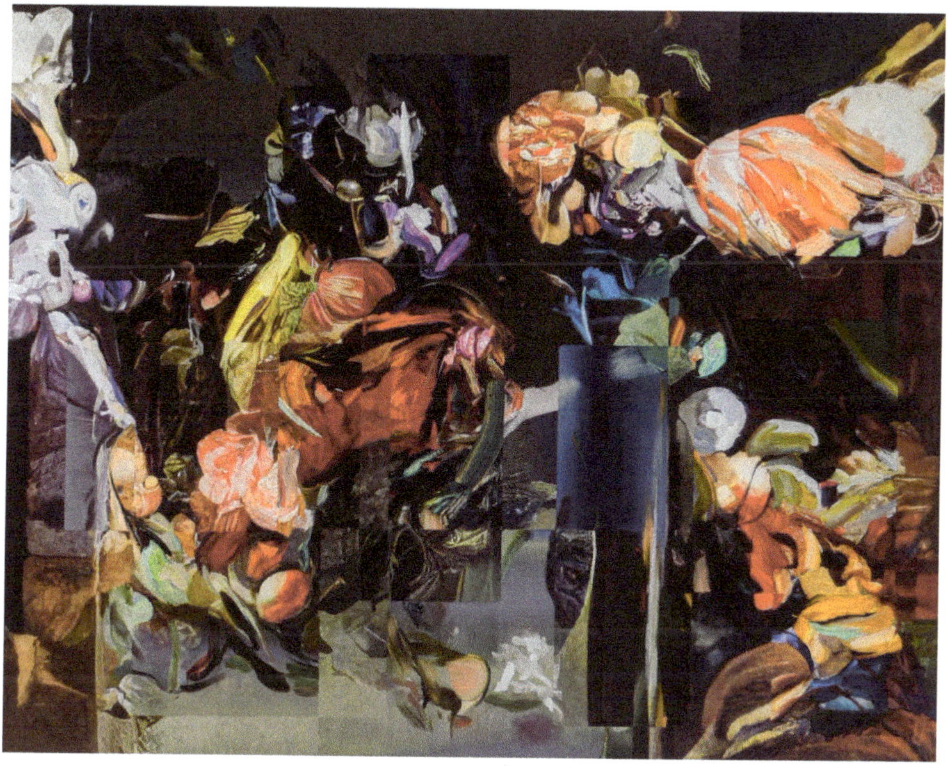

data. In her project "Artificial Natural History (figure 5)[15]," she trains AI on natural imagery to create plausible but impossible lifeforms. using public domain images and various models to generate a natural history that never existed. Her work is speculative, questioning how knowledge is classified and what gets recorded in history.

Dev Valladares, a Google designer, pushes creative boundaries through playful experimentations. He produced a futuristic short film, "Gardeners (image 6)[16]," using Veo3 (Google's AI video generation engine), envisioning a solarpunk future where humans, nature, and technology harmoniously coexist. Valladares curated clips generated by Veo3 to create the final video, noting that the film's realism and quality arise from the model's extensive training on garden and factory-related data.

---

15. Crespo, *Artificial Natural History.*

16. Valladares, *Gardeners.*

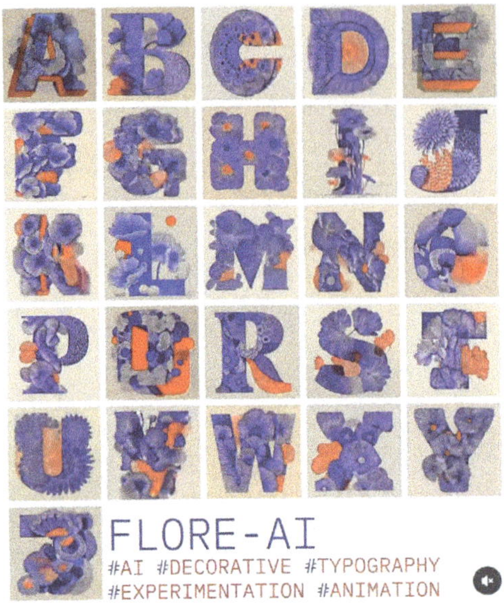

These artists and designers exemplify the evolving relationship between creativity and technology, demonstrating how AI can enhance artistic expression for specific contexts. Their work encourages us to consider the potential of these technologies while reminding us of the unique human touch that continues to drive creativity.

Figure 4. *Flore-AI,* Étienne Mineur.

R3 How does collaborative artmaking with AI redefine the boundaries between human and machine creativity, and how can we identify aspects of creative practice uniquely suited to human ingenuity versus those effectively delegated to AI systems or collaborators?

My review of relevant literature brought me into contact with research on human creativity. Boden's research[17][18] explores how examining AI and machine generalization as they imitate human creativity can enhance our understanding of our own creative processes. The findings provide novel insights into the myth of human creativity by investigating how AI learns to appear to be creative.

As I delved deeper into this topic, it became clear that creativity is not merely a mental process. A formulation in design and art education, emphasizing the complex ability that engages the mind, the heart, and the hands. This ability enables us to generate ideas in our mind, with actions, through physical or digital mediums to execute them, refine them, and repeat again and again, then at some point we stop and deliver the work

17. Margaret A. Boden, *The Creative Mind: Myths And Mechanisms.*

18. Boden, "Creativity and Artificial Intelligence."

to our audience to reflect, to inspire, to make an impact on what we consider important and valuable.

Discussing the factors that impact our sense of creative agency and the value of our work will help us understand why AI could appear to be creative, but not truly be considered a creator of artwork.

One of the questions that AI cannot fully answer is, "Why did you create this artifact?" Regardless of how long the production process takes, it always begins with an input provided by a human, whether that input is a prompt, a question, or a program. AI struggles to recognize and fully comprehend the invisible connections between fragmented tasks and the mental and physical frameworks that guide us toward our vision.

The question of "why" is a key framework that helps us tell our story and build trust and empathy with our audiences. The clearer we are in visualizing the link between the "5 Ws + H" questions—who, what, when, where, why, and how—the greater our level of creative agency becomes. This clarity allows us to convey the unique significance of our work and helps our audience understand and appreciate its value. Ultimately, it answers why our audience should care about what we have created.

This is just one of the missing links that AI fails to comprehend. In our creative agency, the ability to intentionally generate meaningful work arises naturally from the complex, interlocking ecosystems we

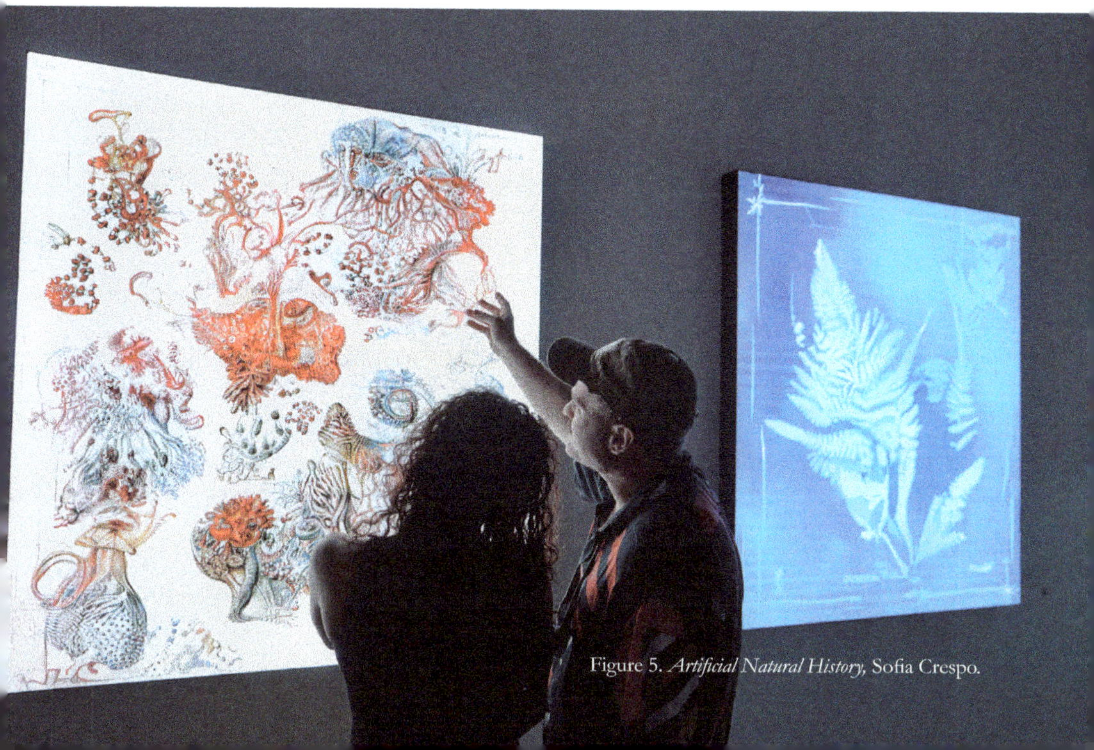

Figure 5. *Artificial Natural History,* Sofia Crespo.

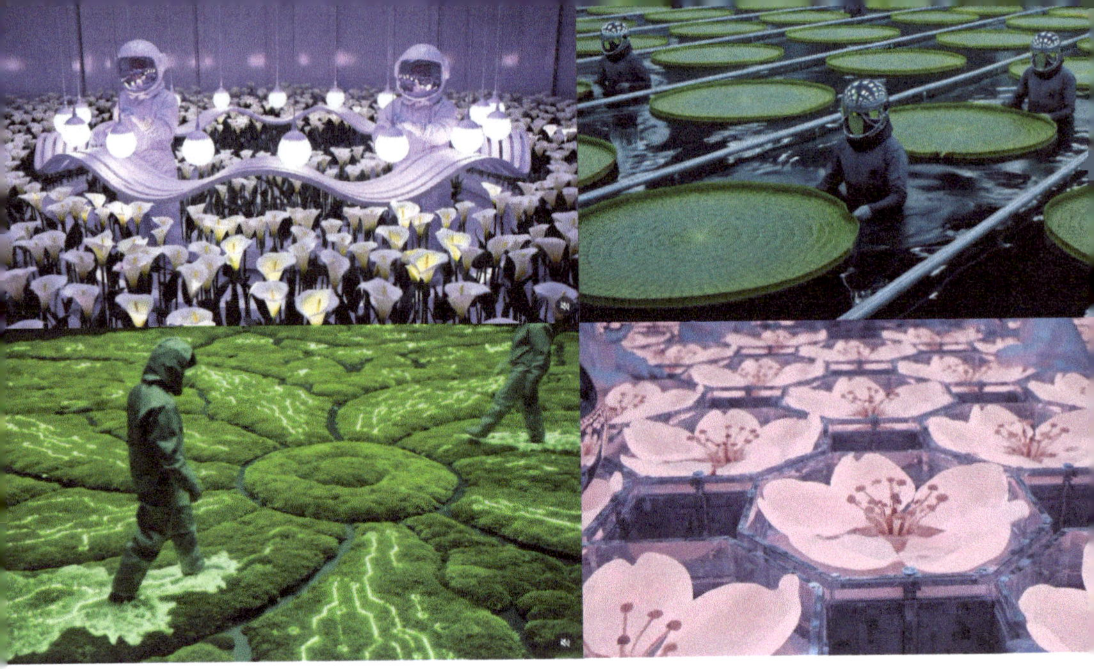

Figure 6. *Gardeners,* Dev Valladares

inhabit. At a minimum, these ecosystems consist of three levels: personal, social, and material. The personal level includes our skills, knowledge, emotions, beliefs, and value systems. We operate based on these value systems, which guide us toward deeper visions and goals. They empower us to confidently make decisions, big and small, throughout our creative production process. The social level involves our relationships and interactions with others, including the influence of communities and networks that challenge and enrich our work. The material level pertains to the resources and tools we have access to, such as technology, materials, and environments. Our creativity arises from a holistic understanding of the limitations, as well as the opportunities these interconnected ecosystems offer, which AI may not fully grasp.

AI struggles to fully grasp the connections between the past, present, and future within these interrelated ecosystems. For artists and designers, the significance of their creative work is often evident when it clearly demonstrates its relationship to past experiences, current events, or future possibilities, emphasizing whether it continues or shifts from their previous creations. This interconnectedness of a body of work can take years or even a lifetime to develop. However, the sense of narrative continuity, an understanding of how past experiences, current context, and future aspirations are intertwined, is something that AI cannot replicate.

We are the storytellers of our own work, shaping how it is presented and delivered, how it is used or applied, and why it matters now. We determine why we are the ones working on it and how it connects to our audience, as well as the past and future of our work. Our AI partners take a long time to understand us, and even then, they cannot fully grasp the social, cultural, and ethical values we hold. While AI can mimic our thinking, it lacks the heart and higher purpose that drive human creativity.

In conclusion, AI excels at rapidly processing and learning from vast amounts of information, but it struggles to comprehend the intricate relationships between data points and the context in which these data points exist. Additionally, AI lacks the ability to deeply engage with the embodied experiences that require a living body, consciousness, and the depth of human emotions and lived experiences. Ultimately, the dialogue between human and machine creativity prompts us to reevaluate our role as storytellers. This dialogue invites us to redefine what it means to imagine, innovate, and inspire.

Bibliography

Adobe Communications Team. Adobe's AI and the Creative Frontier Study Reveals Creators' Views on the Opportunities and Risks of Generative AI. October 8, 2024. https://blog.adobe.com/en/publish/2024/10/08/adobes-ai-creative-frontier-study-reveals-creators-views-opportunities-risks-generative-ai?utm_source=chatgpt.com.

Anantrasirichai, Nantheera, and David Bull. "Artificial Intelligence in the Creative Industries: A Review." *Artificial Intelligence Review* 55, no. 1 (2022): 589–656. https://doi.org/10.1007/s10462-021-10039-7.

Boden, Margaret A. "Creativity and Artificial Intelligence." *Artificial Intelligence* 103, no. 1 (1998): 347–56. https://doi.org/10.1016/S0004-3702(98)00055-1.
Calvert, Tiffany. *464*. 2025. https://www.tiffanycalvert.com/2025.

Crespo, Sofia. *Artificial Natural History*. 2024 2022. Digital. https://sofiacrespo.com/artificial-natural-history/.

DALL·E Now Available without Waitlist. September 28, 2022. https://openai.com/index/dall-e-now-available-without-waitlist/?utm_source=-

chatgpt.com.

David Salle. *Some Versions of Pastoral.* April 10, 2025.

Deckker, Dinesh, and Subhashini Sumanasekara. "A Review of AI-Powered Creativity: The Intersection of AI and the Arts." *International Journal of Global Economic Light (JGEL)* 11, no. 4 (2025). https://doi.org/10.36713/epra0003.

Kelin Carolyn Zhang and Ryan Mather. *Poetry Camera.* 2025. https://poetry.camera/.

Margaret A. Boden. *The Creative Mind: Myths And Mechanisms.* Second. 2004. chrome-extension://efaidnbmnnnibpcajpcglclefindmkaj/https://www.tribuneschoolchd.com/uploads/tms/files/1595167242-the-creative-mind-pdfdrive-com-.pdf.

Mario Klingemann. *'Appropriate Response.'* 2020. https://onkaos.com/work/appropriate-response/?utm_source=chatgpt.com.

Mario Klingemann. *Appropriate Response.* August 29, 2020.

https://quasimondo.com/2020/08/29/appropriate-response/.

Mario Klingemann. *Appropriate Response by Mario Klingemann.* Portfolio. n.d. https://quasimondo.com/2020/08/29/appropriate-response/.

Mineur Etienne. *FlORA-AI.* 2025. Digital. https://www.instagram.com/p/DMAd_YYiK0A/.
Nicholas Davis and Janet Rafner. *AI Drawing Partner: Co-Creative Drawing Agent and Research Platform to Model Co-Creation.* 2025. https://doi.org/arXiv:2501.06607.

Singh, Shubham. *ChatGPT Statistics & Total Users (2025): DAU & MAU Data.* July 24, 2025. https://www.demandsage.com/chatgpt-statistics/?utm_source=chatgpt.com.

Thaddaeus Ropac Exhibition: Some Versions of Pastoral. 2025. https://ropac.net/online-exhibitions/197-david-salle-some-versions-of-pastoral/.

Valladares, Dev. *Gardeners.* 2025. https://www.instagram.com/devethan-valladares/?hl=en.

Wiggers, Kyle. *ChatGPT Users Have Generated over 700M Images since Last Week, OpenAI Says.* April 3, 2025. https://techcrunch.com/2025/04/03/chatgpt-users-have-generated-over-700m-images-since-last-week-openai-says/?utm_source=chatgpt.com.

Wondercraft. *AI in Content Creation 2025, a New Report from Wondercraft, Reveals That over 80% of Creators Are Now Using AI in Their Workflows—Nearly 40% from Start to Finish.* May 14, 2025. https://soundsprofitable.com/press-release/ai-in-content-creation-2025-a-new-report-from-wondercraft-reveals-that-over-80-of-creators-are-now-using-ai-in-their-workflows-nearly-40-from-start-to-finish/?utm_source=chatgpt.com.

Cultural Capital
100 Years of Letters

By Leah Modigliani, Ph.D.

Cultural Capital is a visual installation and/or book comprised of dozens of fictional letters of recommendation for a Guggenheim Memorial Foundation Fellowship. Each letter was written in the appropriated voice and style of living and deceased historic and contemporary critics, curators, historians, and art-world influencers in support of my work. Exceeding the time frame of my own life, the letters are written by such luminaries as Denis Diderot, Charles Baudelaire, John Ruskin, Guillaume Apollinaire, Walter Benjamin, Clement Greenberg, Linda Nochlin, Lucy Lippard, Griselda Pollock, Rosalind Krauss, Jeff Wall, and many others. Each letter reflects what ChatGPT determines the social mores of the original author's time are, including their particular interests. The letters are also strangely similar, as they effusively describe Modigliani's artistic innovation and success through the Western English "jargon of authenticity" (Adorno) that has built and supported the discourse of global contemporary art over a century, including the prejudices and assumptions about greatness embedded therein. This cheeky and often funny work nonetheless speaks seriously about how patterns of recognition move along well-worn paths that are likely to become even more entrenched through the use of AI for writing.

For many decades the John Simon Memorial Guggenheim Foundation used a generic form to solicit recommendations for applicants to the program that they called a "report." I will be using this form, using a typewriter, longhand ink, or a computer to write out the letters composed by ChatGPT. They will hold visual interest and pique curiosity remeniscent of Mel Bochner's piece from his In his 1966 exhibition, "Working Drawings and Other Visible Things on Paper Not Necessarily Meant To Be Viewed as Art" (figure 1).

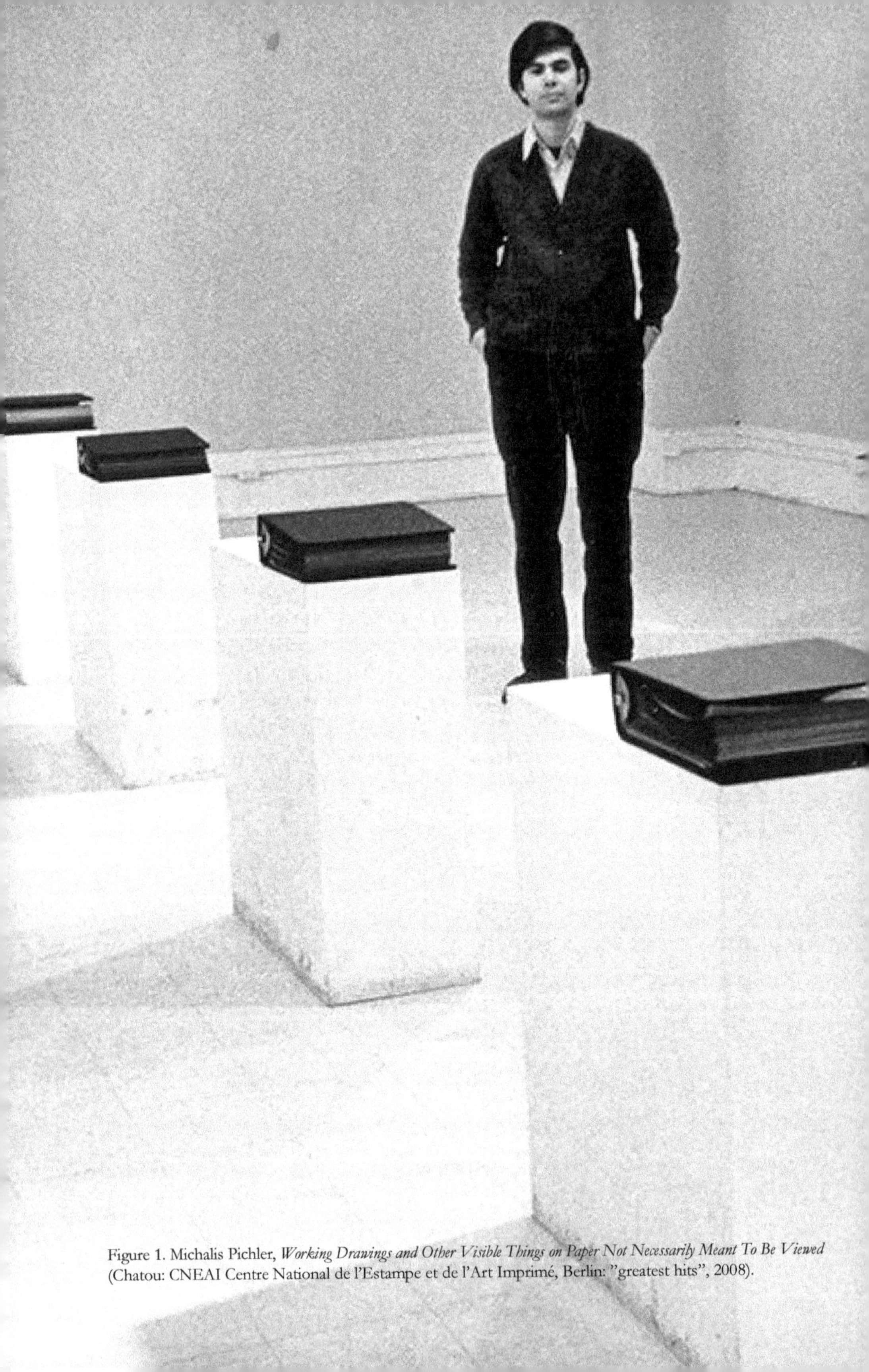

Figure 1. Michalis Pichler, *Working Drawings and Other Visible Things on Paper Not Necessarily Meant To Be Viewed* (Chatou: CNEAI Centre National de l'Estampe et de l'Art Imprimé, Berlin: "greatest hits", 2008).

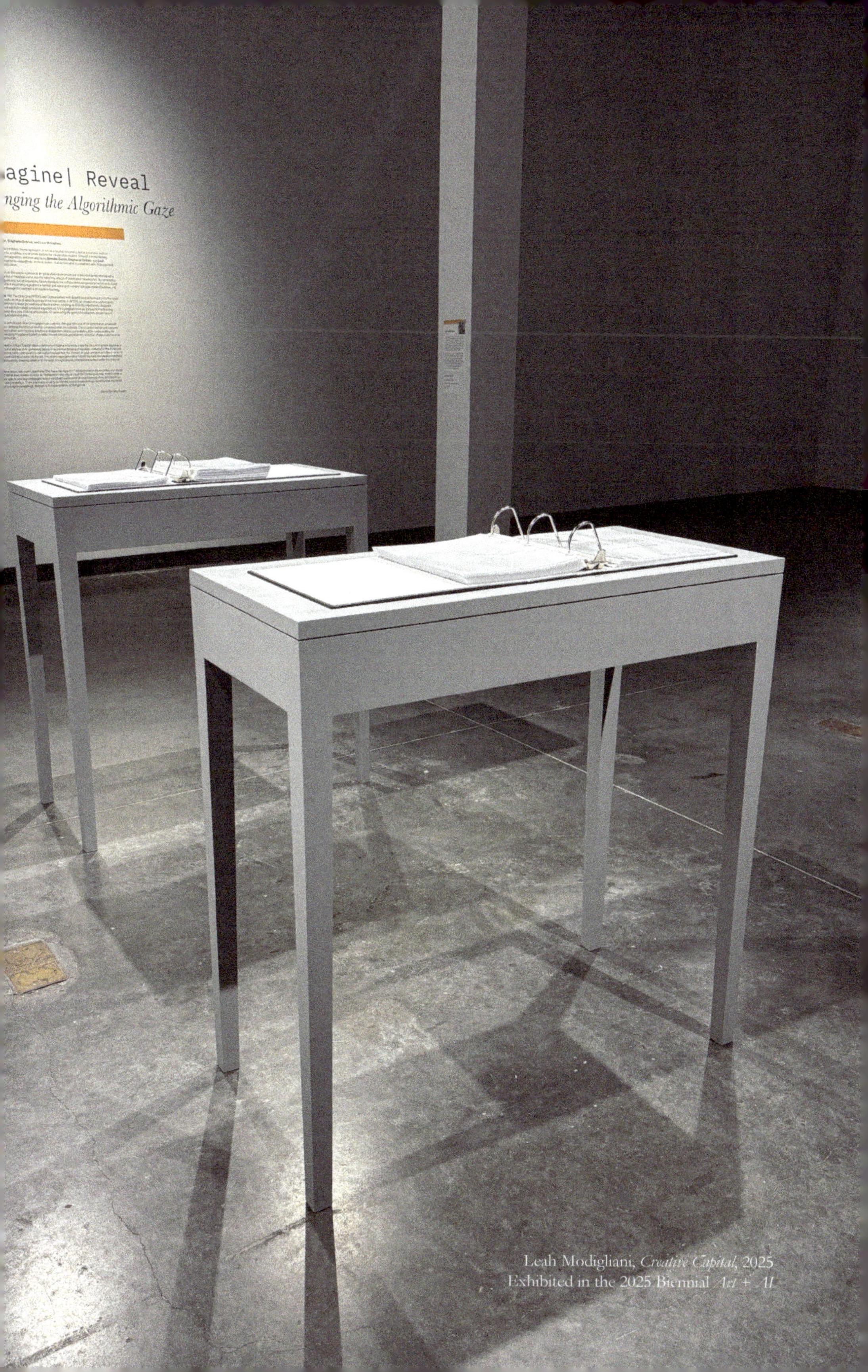

Leah Modigliani, *Creative Capital*, 2025
Exhibited in the 2025 Biennial *Art + AI*

JOHN SIMON GUGGENHEIM MEMORIAL FOUNDATION
Confidential Report on Candidate for Fellowship

FELLOW

W17F

Candidate:

I write in strong support of Leah Modigliani's candidacy for a Guggenheim Fellowship. In a period when the term "modern art" is often diluted through overuse and misapplication, Modigliani's work reminds us that the term ought to be reserved for those practices that pursue form and structure with discipline, coherence, and invention. Modigliani operates squarely within the lineage of advanced modernism—a lineage defined by its insistence on the autonomy of the medium and the critical refinement of visual language. The work demonstrates a commanding grasp of spatial dynamics, material presence, and compositional restraint. It engages problems of volume and architecture not as external subjects, but as internal formal questions to be worked out through the logic of the object itself. There is no reliance here on gimmick or sensationalism. Modigliani's work stands out precisely because it resists the temptation to dilute form with commentary or spectacle. Instead, it returns the viewer to the essential

Signed ______________________ Date ____________

Position or title Clement Greenberg

Address ______________________

(Please return to John Simon Guggenheim Memorial Foundation, 90 Park Avenue, New York, N.Y. 10016, at your earliest convenience. An addressed envelope in enclosed.)

W17F

experience of seeing—of confronting the work on its own terms and discovering in it a rhythm, a tension, a resolution that can only be achieved through visual means. This is not to say the work lacks thought—on the contrary, it is charged with it—but the thought is embedded in structure, in the shaping of material and space, not draped over the work like an explanation. It is rare to see such clarity of intention paired with such precision of execution. The strength of Modigliani's practice lies in its refusal to compromise: it is rigorous, unsentimental, and formally intelligent. These are the qualities that have always defined the best modern art, from its inception in the late nineteenth century through its most vital developments in the present. I recommend this artist without reservation. The work deserves the recognition and support that a Guggenheim Fellowship would rightly confer.

Signed ______________________ Date__________

FELLOW

A36F

JOHN SIMON GUGGENHEIM MEMORIAL FOUNDATION

Confidential Report on Candidate for Fellowship

Candidate:

I am writing to support Leah Modigliani's application for a Guggenheim Fellowship. Her work is part of an urgently needed redefinition of what modern art can do—historicaly grounded, politically engaged, and visually intelligent without being formalist for its own sake.

Modigliani works across sculpture, installation, photography, and text, but her practice is unified by a consistent challenge to dominant narratives in art and history. Her investigations into labor, protest, authorship, and the politics of space echo the critical aims of conceptualism but are sharpened by a deep historical awareness. This is work that understands where it comes from and asks hard questions about where it's going.

Unlike artists who merely illustrate theory, Modigliani embeds critique in material form. Her work doesn't separate aesthetics from ethics, or form from content—it recognizes that all are part of the same cultural system. She refuses the neutrality that institutions still often expect from artists, and instead creates work that speaks back—calmly, structurally, and with precision.

She's also a sharp writer and thinker, which informs and strengthens her visual work. That kind of hybrid practice—both intellectually grounded and materially engaged—has always been rare, and it's even more crucial now.

Modigliani is doing the kind of work that makes space for others. A Guggenheim would not only support her—it would support the kind of critical, meaningful art practice we need more of.

Sincerely,
Lucy Lippard

Signed ______________________ Date ____________

Position or title ______________________

Address ______________________

(Please return to John Simon Guggenheim Memorial Foundation, 90 Park Avenue, New York, N.Y. 10016, at your earliest convenience. An addressed envelope in enclosed.)

A27f

JOHN SIMON GUGGENHEIM MEMORIAL FOUNDATION

Confidential Report on Candidate for Fellowship

FELLOW

Candidate:

To the Guggenheim Fellowship Selection Committee,

It is with great enthusiasm that I write in support of Leah Modigliani's application for the Guggenheim Fellowship. Modigliani is an artist whose work opens up new and urgent conversations—about history, resistance, visibility, and the infrastructures that shape the making and reception of art. Her practice is a laboratory for forms of knowledge and unlearning, always asking: how can we think otherwise?

I first encountered Modigliani's work through her incisive combinations of sculpture, photography, and text—always shifting, never static. Hers is a practice that does not rest in medium, but moves across architecture, urbanism, performance, politics. As Édouard Glissant might say, it is a practice of relation: connecting geographies, memories, and discourses that don't always get to meet.

Modigliani is not only an artist but a researcher, a historian, a storyteller. She belongs to that group of artists I admire most—those who create ecosystems of thought. Her work proposes alternate readings of modernism and protest, authorship and labor. It is always precise, but never closed. It invites dialogue. It insists on complexity.

In my conversations with artists over the years, I've come to believe the most enduring practices are those with long duration—those that unfold over time, refusing easy answers. Modigliani's is one of these. Her project is not bound to trend or market, but to a deep inquiry into the conditions under which art and history are shaped.

She asks what it means to remember. To reframe. To reclaim space. These questions are not only central to art—they are central to how we live now.

The Guggenheim Fellowship would be a vital support for this ongoing work. Modigliani's voice is one we need—rigorous, poetic, and profoundly committed.

With best wishes,
Hans Ulrich Obrist

Signed ______________________ Date ____________

Position or title Artistic Director Serpentine Galleries

Address Kensington Gardens, London W2 3XA

(Please return to John Simon Guggenheim Memorial Foundation, 90 Park Avenue, New York, N.Y. 10016, at your earliest convenience. An addressed envelope in enclosed.)

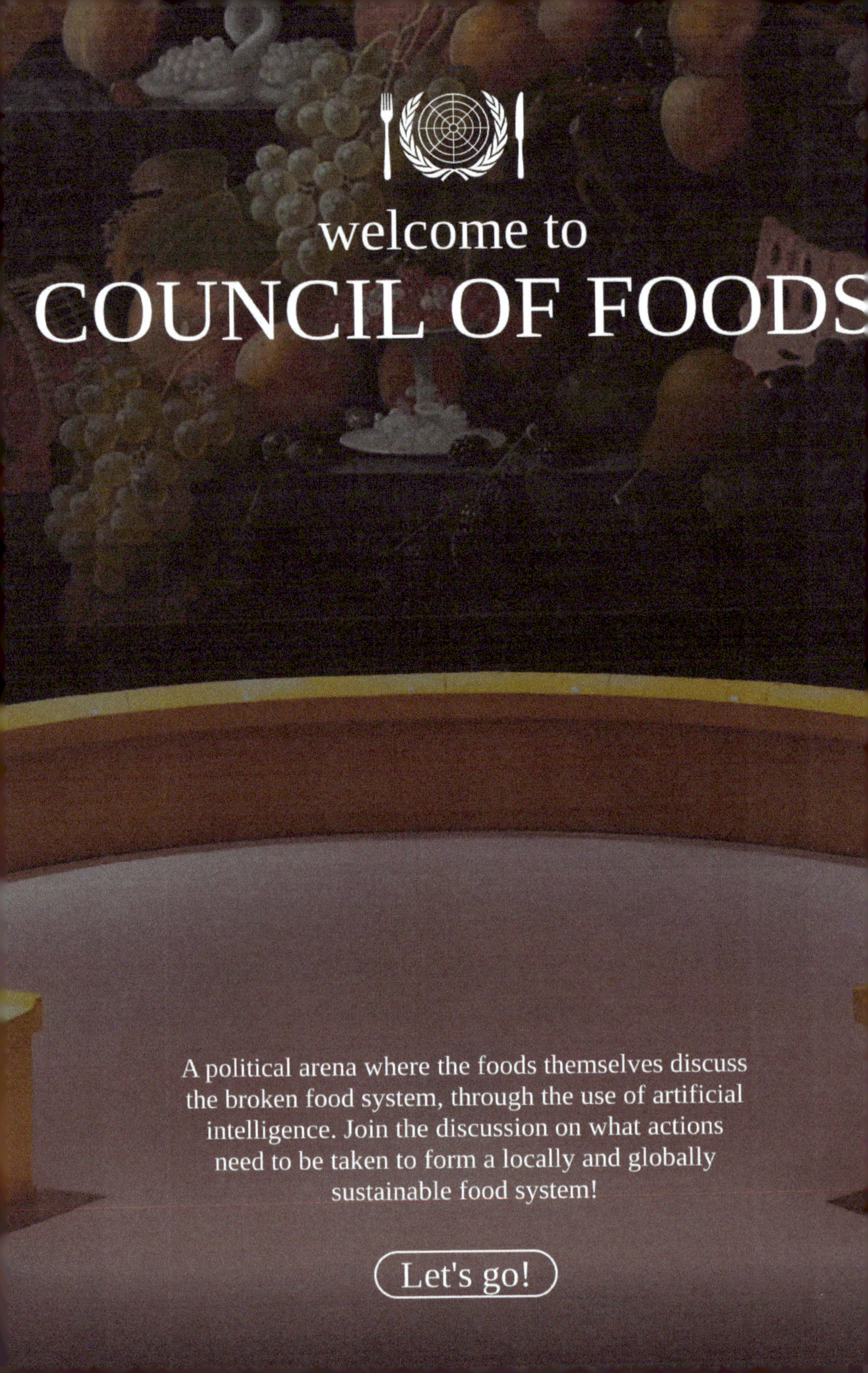
welcome to
COUNCIL OF FOODS
A political arena where the foods themselves discuss the broken food system, through the use of artificial intelligence. Join the discussion on what actions need to be taken to form a locally and globally sustainable food system!
Let's go!

The Council of Foods

A Case Study of Becoming and the Rhizome in Artificial Intelligence

By Xiaoxu Zhang

Nonhuman Nonsense, an artist collective dedicated to change the relationship with the non-human, embracing paradoxical stories that open the public imaginary (S+T+ARTS, 2025). In the AI interactive artwork *The Council of Foods* (Nonhuman Nonsense, 2025), they liberate themselves from the traditional role of the artist and, collaborate with an Artificial Intelligence system as a programmer to construct the algorithmic framework and implement the generative logic. In this project, audiences are invited to select *The Issue* that will be discussed in the council and *The food* that will participate in it the council. They also take part in this virtual conference with the topic of environmental protection. The AI system responds to the audience's comments in real time, generating personified statements from the food representatives—voicing anxieties and reflections on the ecological crisis.

> *The prototype Council of Foods demonstrates technological readiness for personalized learning experiences through AI and machine learning, with council members (in this case prompted foods) potentially acting as tutors, as discussion partners, as criticizers or as best practice experts in dialogue with individual policy makers or groups of policy makers* (S+T+ARTS, 2025). (Figure 1)

Within this space of multiple agencies, a question arises: who is the true author of the artwork? Is it the Nonhuman Nonsense who designed the framework, the audience who select the issue and the food, or the AI that executes the computational processes? And who is telling the story? Is it the algorithmic model behind the artwork, or the food avatars that have been given a voice? The boundaries between agents become blurred, and the binary roles of the artist and the audience are reconfigured.

The charm of AI art lies precisely in its refusal to be rigid, always in a dynamic process of becoming, connection, and lines of flight (Deleuze & Guattari, 1987). The traditional single-line arts create logic of *artist-artwork-audience* is broken with the participation of AI, and instead evolves into a polycentric, boundaryless mode of creation and experience (Nadine, 2016). As Deleuze and Guattari proposed in *A Thousand Plateaus* through the concepts of *becoming* and *rhizome*, *The Council of Foods* exemplifies a *non-hierarchical, open-ended rhizomatic system* driven by AI, where fluidity redefine authorship (Deleuze & Guattari, 1987). At the same time, AI acts as the dynamic node in the artwork's creative network to change the existing viewing relationships and enables the audience to transforming audiences from passive spectators into co-creators and triggers within this rhizomatic generative process.

Under the creative background of post-humanism constructed by *Nonhuman Nonsense*, with the theoretical framework of Deleuze and Guatali, this article will explore whether the intervention of AI breaks the fixed identities of *artist-artwork-audience*, redistributes artistic authority, and challenges traditional definitions of "art," thereby triggering a new round of revolution in avant-garde art.

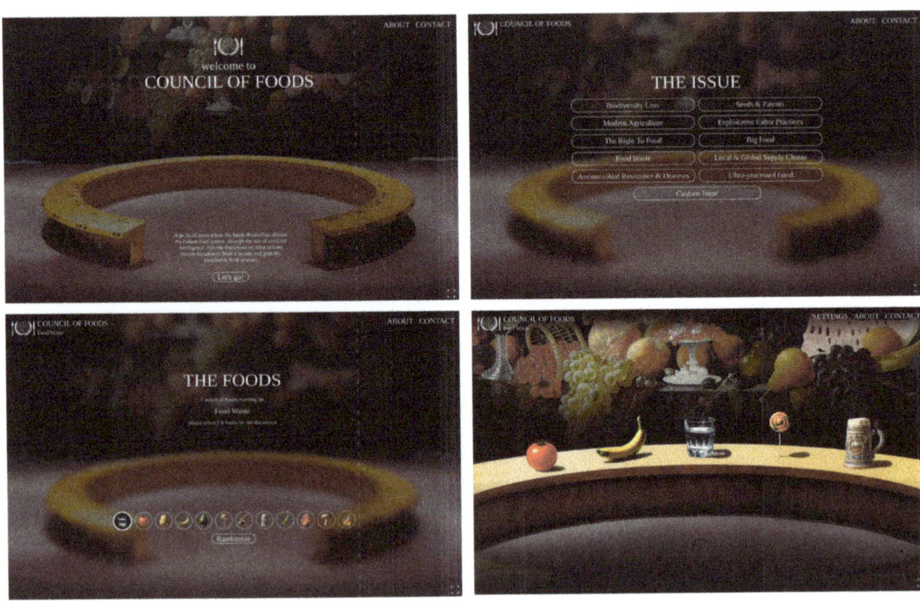

Figure 1. The process of initiating *The Council of Foods*

Shared Authorship in the Artistic Practice of Nonhuman Nonsense

While traditional humanism places "human" at the center, in posthumanism, nonhuman agencies such as technology, animals, and the environment become equally important components that shape the world we live in. Posthumanism aims at overcoming anthropocentrism (Braidotti, 2013), which means that in the context of posthumanism, humans have lost their original priorities and no longer appear as protagonists, subjects, or manipulators. Posthumanism values non-human or a-personal life (Braidotti, 2013), where animals, the environment, and technology also possess their own rights, emotions, and responsibilities. By this means, people should rethink links of affectivity and responsibility not only for non-anthropomorphic organic others (Braidotti, 2013), and reconstruct those technologically mediated, newly patented creatures we are sharing our planet with (Braidotti, 2013). This mode of rethinking and reconstruction corresponds to the creative intent of *Nonhuman Nonsense*.

Nonhuman Nonsense is an interdisciplinary art collective centered around the philosophy of posthumanism. Their works transcend the boundaries of art, ecological philosophy, and artificial intelligence. Their official website bluntly separating the human from the nonhuman is nonsense (Nonhuman Nonsense, 2025). As an artist collective that creates in the context of posthumanism, the artworks of *Nonhuman Nonsense* are filled with nonhuman entities and emotionless creative subjects and centers. At the practical level, in their artworks, animals, plants, objects and ecological techniques have equal say and creative rights (Baart, 2021).

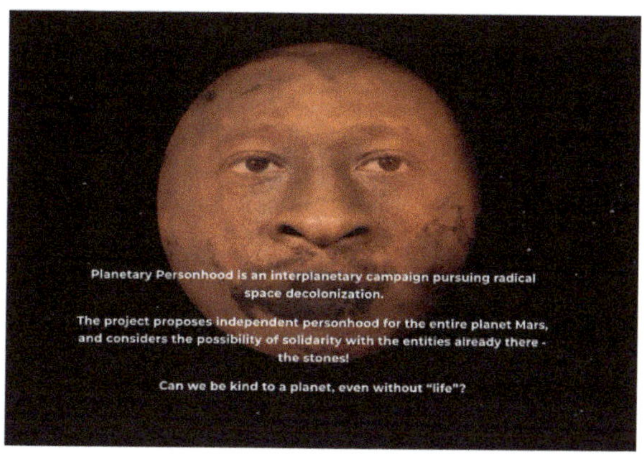

Figure 2. Planetary personhood by *Nonhuman Nonsense*

Through a series of cross-species or cross-entity art projects, such as a declaration that endows Martian rocks with Planetary Personhood and the genetically stained

Pink Chicken Project, Nonhuman Nonsense is dedicated to helping the audience establish connections with non-human and unknowable things, find the intersection between the human world and the non-human world, and stop counting, classifying, categorizing, and sorting non-humans.

Decentralized Authorship

In *The Council of Foods*, the audience, AI and The Food have engaged in a dialogue together, constructing a nonlinear, multi-centered "rhizomatic" equal creative network in the context of posthumanism. Like many AI-driven artworks, artists emerge as "system architects" in The Council of Foods (Yike& Qifeng, 2022). They no longer produce the artwork with predetermined artistic presentation but instead focus on designing an interactive framework that governs the generative process. However, they still maintain the order of the framework, for example, artists will address the algorithmic biases that AI may cause, such as eliminating generated content containing racist metaphors. They have the right to influence the presentation of the works, but they cannot truly determine the ultimate direction of the works.

As the driving engine in the artworks, although there are differences in how ideas are generated, no qualitative difference between AI and human-generated creativity (Jennifer & Paul, 2023). They can provide inspiration and creativity for both artists and audiences, and create artworks entirely of its own accord (Michael & Sheena, 2023). However, due to the dialogues and perspectives generated by AI are inherently shaped by its training data, which influences how AI interprets and articulates the conditions of plants, animals, and their hypothetical stances. For instance, the *Irish Potato*, whose narrative background includes the historical trauma of the Great Famine (HISTORY.com Editors, 2025), consistently foregrounds themes such as "poverty" and "colonialism" in its outputs. The real-time input from the audience will also affect and restrict the content created by AI.

In *The Council of Foods*, the Food, like avocado, lollipop, bean, as symbolic entities personified by algorithms, function as creative and representational agents. Within the framework of the artwork, these entities possess their own backgrounds, personalities, and images, and are capable of outputting critical and diverse viewpoints. They can discuss and judge each other's viewpoints and respond to and discuss the questions and viewpoints raised by the audience, which is a manifestation of their creative rights.

AI allows for generative processes creating an artwork that is not only unique but can be influenced by its audience (Nailman, 2025). At *The Council of Foods*, audience as the organizers of this meeting, have the right to change the direction of the meeting by choosing The Issue and The Food, and can participate in the discussions. They are the triggers of the entire meeting and part of the creation.

Artists design rule-based texts or algorithmic frameworks, while audiences and other actors collaboratively input parameters. AI, audiences, and food then generate real-time feedback through experiential participation, enabling a co-creation process where every node—whether human or nonhuman—possesses equal discursive and creative agency.

"Becoming" and "Rhizome"

Becoming is not to imitate or identify with something or someone. Becoming emphasizes a non-teleological, non-linear dynamic process. It is to extract particles between which one establishes the relations of movement and rest, speed and slowness (Deleuze & Guattari, 1987). Becoming emphasizes a non-teleological, non-linear dynamic process. It is not an evolution from a fixed starting point to an endpoint but a continuous production of difference through "passing between" (Deleuze & Guattari, 1983). In *Anti-Oedipus*, becoming is described as "the encounter and interaction of forces", which does not seek unity or stability but constitutes "a movement of escaping identity" (Massumi, 1992), and the essence of this process lies in the emergence of difference. For example, the artwork generated by large language models (LLMs) is entirely derived from the background knowledge it has learned, but its output is not a simple imitation. Instead, they create new forms through difference and disrupt stereotypical patterns via "the repetition of imperfect appearances" (Deleuze, 1994).

The rhizome is Deleuze's subversion of traditional arborescent structures which process the hierarchical and centralized systems. The rhizome,

> *Unlike a structure, which is defined by a set of points and positions, with binary relations between the points and biunivocal relationships between the positions, the rhizome is made only of lines: lines of*

segmentary and stratification as its dimensions, and the line of flight or deterritorialization as the maximum dimension after which the multiplicity undergoes metamorphosis, changes in nature (Deleuze & Guattari, 1987).

such as grassroots networks or the Web2.0 structure of the internet. The characteristics of the rhizome include decentralization, multiplicity, and the potential for rupture and reorganization at any moment. The relationship between becoming and the rhizome is symbiotic, the rhizome as the field of becoming, provides "lines of flight" through its non-hierarchical network, enabling differences to emerge. At the same time, the process of becoming reshapes the rhizome itself. Each act of becoming involves deterritorialization and reterritorialization of existing structures, generating new rhizomatic configurations in the process (Deleuze & Guattari, 1987).

The Council of Foods and "Becoming"

In The Council of Foods, the theory of *becoming* is not an abstract concept; rather, it materializes through AI-empowered nonhuman entities, dynamic audience participation, and unpredictable ethical debates, directly challenges the fixation of identity and the centralization of creative authority, thereby advocating for a shift toward generative artistic practices.

In traditional art, food is merely a symbol or material object—such as the apple in a still-life painting. However, *The Council of Foods* employs AI to transform food into emergent representational and political subjects, a process that embodies Deleuze's "becoming-other" (Deleuze & Guattari, 1987) and food ceases to be a consumable object like the conventional value system. During dialogue process, food characters often engage in dialectical debates. For instance, when discussing the waste of water resources in modern agriculture, Avocado praised the sustainability of small-scale agriculture, while also suggesting that small-scale agriculture needs to enhance productivity and efficiency. This paradoxical and dialectical stance aligns with the essence of becoming, which rejects identity-based logic and embraces the fluidity of intensities (Deleuze & Guattari, 1987).

In traditional art, audiences function as passive recipients or evaluators of meaning (Lynne, 2016). In *The Council of Foods*, audiences become triggers and co-creators in the generative process through real-time interaction. They actively participate in and influence the outcomes of artwork. Within this artwork, audiences select discussion issues, choose participating foods, and ask questions or express their commons during meeting. Their behavior mirrors Deleuze's concept of the *nomad*, they locally perturbing the system, triggering AI to configure the ethical framework of food roles, and changes the direction of the conversation. This transformation of identity and acquisition of rights break the passive identity of the audience as the other, allowing the audience to obtain the right to express their desires, driving the artistic process to generate in the unknown direction, and ultimately bringing about a wide variety of presentation results for art.

In *The Council of Foods*, the character of AI is not a passive tool but an engine and participant in artwork creative process, it assumes multiple, uncontrollable roles while internally coordinating and manipulating outputs. AI equippe with advanced learning, dialogic, and logical capabilities (Floridi, 2019), Although the initial parameters of the AI model are set by the artist, such as the background, personality and ethical tendencies of the food characters, its output is influenced by its self-learning and the audience's viewpoints, and the generated results often overflow the preset framework. For instance, AI might relate the remarks of mushrooms to colonial history beyond mere ecological issues. This overflow is the essence of generation - the unpredictability of differences, which are brought about jointly by artists, AI and audiences, and ultimately presented through the dialogue of food.

The Fluid Identity in The Rhizome

Becoming is the driving force of the rhizomatic network, while the rhizome serves as the topological field in which becoming unfolds (Deleuze & Guattari, 1987). Under the driving force of becoming, artists, AI, audiences, and food representatives form a non-hierarchical, open-ended identity network characterized by fluidity and decentralized agency. As mentioned above, in the post-human context, artists, audiences, Ai, and food all have equal authorship. During the entire process of generating the artwork, they cannot solely control the direction of the artwork. For example, although the initial parameters of the AI language model are set by the artist, its output is affected by the interaction of the audience and the dynamics of the data, forming a constantly changing relationship network.

Within this network, the identities of all co-creative nodes are in constant flux. Taking the audience as an example, when the meeting began and the food characters expressed their opinions, the audience listened to the content about environmental protection, they are the audience. And when audiences participate in the council by expressing their opinions, they transform into skeptics or evaluators.

In this ever-expanding, intricately entangled rhizome, the audience's identity is no longer fixed but fluidly evolves alongside the generative process. Similarly, the roles of AI, artists, and food representatives shift and transform within the rhizomatic structure. Council of Foods converts art from a *human monologue* into a *multispecies dialogue* through its rhizomatic networks and processes of becoming. This practice not only challenges traditional artistic paradigms but also implies a posthuman ethics, in a period of ecological crisis and technological singularity, art must become a rhizome that accommodates nonhumans, blurs boundaries, and sustains perpetual generation—a truly democratic parliament.

AI Arts: Where Exactly Are the Limits of What Could Be Considered Art?

Traditional artworks are confined to a closed creative process centered on the triad of "artist-artwork-audience"—even in participatory art, audiences rarely engage in or influence the entire creative process, let alone reciprocally reshape the artwork itself (Bishop, 2012). However, the AI Arts represented by *The Council of Foods* is no longer a closed artwork system, but an open experimental process. Every parliamentary discussion is an ethical exercise, and its value lies in stimulating reflection. From the content level, it inspires the audience to think about food from the perspective of non-human character food. From a technical perspective, it evokes audiences to think when AI is involved in the creation of art, what exactly changes this creative process.

The rhizomatic network in AI art, with its inherent unpredictability, invites continual audience engagement. The relentless generation and learning of AI, food representatives' persistent advocacy for ecological issues, and artists' diversified presentations collectively forge a new ethical experimentation in both arts create way and content as reflected by Ruth:

Before Duchamp's ready-mades, the prevailing ideas about what constituted art were very traditional. Art was defined largely in terms of its aesthetic properties: value was derived from technical proficiency of the artist, and, loosely, the way it conveyed emotion (Ruth 2023).

Today, a growing number of AI-driven artworks are redefining the creative process by granting participants agency over authorship, thereby transforming traditional artistic relationships and modes of spectatorship. These works provoke critical debates on questions such as *who is the true creator? Does AI claim to authorshi*p? *intellectual property*, and t*he boundaries of art*...... As Deleuze asserted, becoming is not about attaining a state but about escaping the present (Deleuze & Guattari, 1987). When AI stops functioning merely as a tool and audiences transcend passive observation, the very ontology of art is reconfigured: How is the artwork presented to the audience? Do the audience's experiential engagement and the process of co-creation become the artwork's most essential components?

Conclusion

This paper examines how *The Council of Foods* by artist collective *Nonhuman Nonsense* employs AI as co-creator to blurs the fixed identities and authorship structures of traditional art, driving art toward a transformative shift into generative praxis. In this process, AI not as a tool or technology but as the co-creative force, generating content that frequently exceeds preset frameworks and manifests the unpredictability of difference. Rooted in the decentralized architecture of rhizomatic networks, this creative model creates spaces for the emergence of difference while reconfiguring audiences and other non-human agents from passive recipients into co-creators.

The Council of Foods challenges conventional artistic paradigms and proposes a posthuman ethics, art will evolve into a rhizomatic structure that accommodates nonhumans, blurs boundaries, and sustains perpetual generation. This practice reveals the potential of AI art and provides a new direction for future artistic creation.

Bibliography

Baart, R. (2021, February 1). *Talking non-human nonsense with Nonhuman Nonsense*. Next Nature. https://nextnature.org/en/magazine/story/2020/nonhuman-nonsense

Bishop, C. (2012). *Artificial hells: Participatory art and the politics of spectatorship*. Verso Books.

Braidotti, R. (2013). *The posthuman*. Polity Press.

Conner, L. (2016). Taking back the arts: 21st century audiences, participatory culture and the end of passive spectatorship. *L'Ordinaire Des Amériques, 220*. https://doi.org/10.4000/orda.2773

Deleuze, G., & Guattari, F. (1983). *Anti-Oedipus: Capitalism and schizophrenia* (R. Hurley, M. Seem, & H. R. Lane, Trans.). University of Minnesota Press. (Original work published 1972)

Deleuze, G., & Guattari, F. (1987). *A thousand plateaus: Capitalism and schizophrenia*. University of Minnesota Press.

Haase, J., & Paul. (2023). *Artificial muses: Generative Artificial Intelligence chatbots have risen to human-level creativity*. https://doi.org/10.48550/arxiv.2303.12003

History.com Editors. (2017, October 17). *Irish Potato Famine: Date, Cause & Great Hunger | HISTORY*. HISTORY. https://www.history.com/articles/irish-potato-famine

Horton Jr, C. B., White, M. W., & Iyengar, S. S. (2023). Bias against AI art can enhance perceptions of human creativity. *Scientific Reports, 13*, Article 19001. https://doi.org/10.1038/s41598-023-45202-3

Nonhuman Nonsense. (2025). *Council of Foods*. Council-of-Foods.com. https://council-of-foods.com/foods

S+T+ARTS. (n.d.). *The Council of Foods*. https://starts.eu/hungryecocities/the-council-of-foods/

Schröder, N. J. (2018). The lens of polycentricity: Identifying polycentric governance systems illustrated through examples from the field of water governance. *Environmental Policy and Governance, 28*(4), 236–251. https://doi.org/10.1002/eet.1812

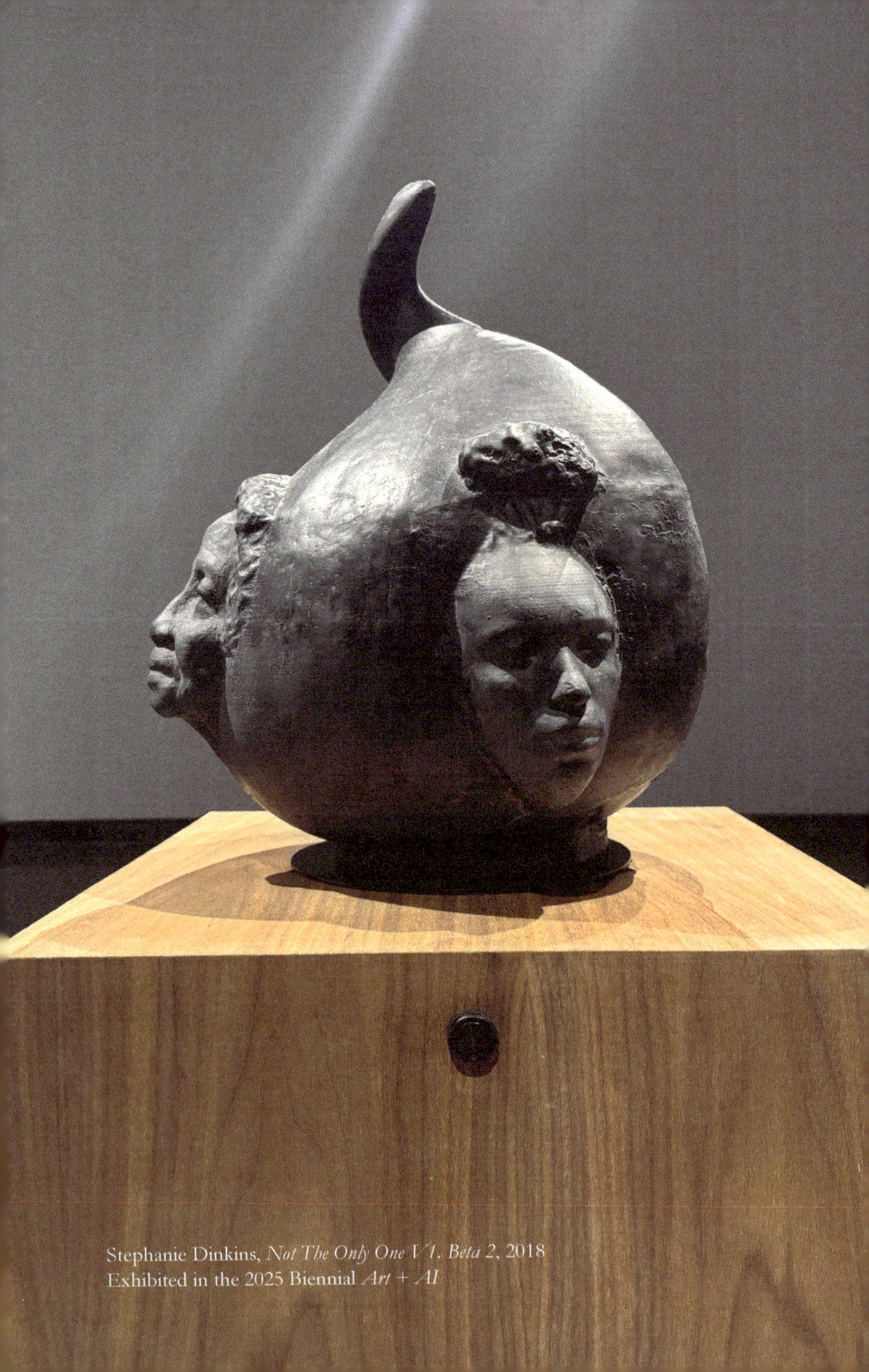

Stephanie Dinkins, *Not The Only One V1. Beta 2*, 2018
Exhibited in the 2025 Biennial *Art + AI*

About
The Authors

In alphabetical order.

Marnie Benney
Marnie Benney is an Independent Curator exploring the intersection of contemporary art, technology and design. Over the past decade, she has produced more than 40 exhibitions in city centers, public spaces, galleries and festivals worldwide. Her work has been showcased at leading institutions including the New York Hall of Science, University of Cambridge in London, the National Aquarium in Maryland, and The Misk Art Institute in Riyadh, Saudi Arabia. Benney's work investigates the societal, cultural and future implications of technology through the lens of contemporary art. In 2019, she founded AIArtists.org, the world's largest community of artists working with Artificial Intelligence, where she serves as Curator and provides a platform for artists to explore the future implications of AI on society.

Mark Burchick
Mark Burchick is a Capital Emmy-award winning filmmaker and interdisciplinary artist from Baltimore, Maryland. He holds an MFA in Studio Arts from Towson University. Sourcing from academic fields such as religious and folklore studies, anthropology, and psychology, his current body of work involves the study of belief and the roles that media and technology play in influencing our beliefs. These conceptually driven single channel videos and interactive multimedia installations invite viewers to question their perceptions and examine how belief systems have structured our world.

Radu-Florentin Diaconu, Ph.D
Dr. Diaconu is a discourse analyst and content editor. He holds a Ph.D. in Linguistics from the "Alexandru Ioan Cuza University" of Iași, where his research focused on presidential communication through a socio-rhetorical lens. He is currently Managing Editor at AmpiFire, where he

oversees editorial strategy, writer training, and content quality in an AI-integrated workflow for business content development. His research interests include AI and media studies, critical AI literacy, rhetorical and discourse analysis, and the cultural impact of automation in communication and content production.

Amy Hicks

Amy Hicks is a Philadelphia-based artist, educator, and member of the artist collective Grizzly Grizzly. Through image processing and purposely rudimentary erasure-animation she combines pre-existing material with original footage to develop projects that question belief systems and the impact they have on the individual and social body. Her work has exhibited in museums, galleries, and film festivals internationally from California to Croatia. Awards include grants from the San Francisco Art Commission and Film Arts Foundation, a Djerassi Artist in Residence, a Flaherty Film Seminar Fellowship from the Philadelphia Foundation, and was an artist in residence at Interlude Artist Residency (Hudson, NY). She co-edited Grizzly Grizzly's first publication "In Retrospect: 10 Years, 10 Essays, 10 Exhibits" which was supported by the Velocity Fund, a regranting initiative of the Andy Warhol Foundation. She received her MFA from Stanford University and is currently an Associate Professor in the Department of Art and Design at the University of Delaware.

Morgan Joseph Hamilton, Ph.D.

Morgan Joseph Hamilton, Ph.D., is the Gretchen Hupfel Curator of Contemporary Art at The Delaware Contemporary in Wilmington, Delaware. He is dedicated to edu-curation practices and integrating on-site and on-line experiences to broaden accessibility to the arts. He is a proud alum of the Museum Education and Visitor-Centered Curation program at Florida State University. His dissertation, Certain Professionals in an Uncertain Field during Uncertain Times (2024), studied the effects of digital programming on art museum educators' experiences during the crisis. His research interests include digital philosophies, new media and networked art in the contemporary art space, virtual objects and reproduction, and implications of artificial intelligence for art and museums. He wrote a chapter in Rowman & Littlefield's Dimensions of Curation (2023) "Toward an Interactive Model of Competing Values: From Visualization to Toolkit " that emphasized the need for practical, interactive visualizations of highly conceptual theoretical models. His chapter introduced three ways to visualize the complex three dimensions of curation to help curators, educators, and museum professionals plan an intentional exhibition with the audience at its center.

Erica Zoe Loustau

Erica Loustau is an artist who specializes in creating mixed-media sculptural installations. Her works are characterized by the accumulation of simple repeated shapes on geometrically arranged lines, which create a unique experience of a floating mass that resembles a flock or swarm. The obsessive and excessive construction of her installations suggests an attempt to achieve a sublime level of illusion and metaphor, making her art truly remarkable.

Hyeyoung Maeng, Ph.D.

Young (Hyeyoung) Maeng is a contemporary artist and researcher whose work explores the intersections of artificial intelligence, contemporary painting, and posthuman philosophy. She is Assistant Professor of Art at California State University, Fresno, where her teaching and practice focus on expanded painting and drawing, including Korean painting, AI-integrated painting, and critical AI literacy. She received her PhD in Contemporary Arts from Lancaster University, UK, with practice-based research on Deleuze's transcendental realism, and holds MFA degrees from the San Francisco Art Institute and Kyung Hee University, Seoul. Young has presented 16 solo exhibitions and over 70 group exhibitions internationally. She is the author of Art Philosophy after AI (forthcoming, Communication Books, 2025), and her recent publications in SCOPUS- and KCI-indexed journals focus on Deleuze's aesthetics, AI, and artistic creation. Supported by a CSU P-RSCA Impact Grant, her current research explores how human–AI collaboration opens new artistic potentialities of existence and subjectivity in painting.

Leah Modigliani, Ph.D.

Leah Modigliani is Associate Professor of Visual Studies at Tyler School of Art and Architecture. She is an artist and scholar with transdisciplinary engagements in fine arts, art history, critical geography, urban studies, and politics. Modigliani's work represents the liberatory potential (right to the city) and neoliberal revanchism (displacement, punitive laws) of urban experience. She has dwelled upon eviction ("How long can we tolerate this? An incomplete record from 1933-1999," 2016-17); cities destroyed by war ("The City in Her Desolation," 2017) and natural disasters ("Cities of God" series 2021-22), and protests against injustices enacted in cities ("Washington D.C., 1939; Basel, 1957; Berkeley, 1969; Chicago, 1969; London, 1969; Windsor, 1982...," 2015-2018). While often sculptural, her work increasingly cites the form and history of photography, especially photography's role in constituting and deconstructing historical narratives online and in physical archives. Modigliani's visual artworks are complemented and informed by her academic writing about photography and landscape.

Tricky Walsh

is an artist living and working between Melbourne, Australia, and Virginia in the US. Walsh has been awarded a Qantas Foundation Art Encouragement award and won the 2009 Hobart Art prize for the sculpture The Wasp project, has been commissioned to make works for the ICA in Richmond, VA, USA, the 2013 Mona Foma festival, organised by the Museum of Old and New Art, the Museo Madre in Naples, Centre Pompidou in Paris, the Mao Jihong foundation in Chengdu, Monash University Museum of Art, the Tasmanian Museum and Art Gallery and the privately funded Detached Cultural Organisation, and The Old Mercury Building in Tasmania.

Xiaoxu Zhang

Xiaoxu Zhang is the Ph.D. student in Museum Education and Visitor-Centered Exhibitions at Florida State University. She earned B.A. in Lighting Design from the Communication University of China and M.A. in Arts Management from the University at Buffalo. Before beginning her doctoral studies, she worked as an editor responsible for book promotion at Modern Education Press, China Publishing Group. Xiaoxu's research focuses on enhancing visitor engagement in small- and medium-sized museums. She is particularly interested in AI–human collaborative curation and in fostering a sense of belonging within museum spaces. Her current goal is to advance how audiences experience and learn in museums.

Artifice,

noun

The use of a clever trick, or something intended to deceive.

Cambridge English Dictionary

www.ingramcontent.com/pod-product-compliance
Ingram Content Group UK Ltd.
Pitfield, Milton Keynes, MK11 3LW, UK
UKHW022041041026
15591UKWH00013B/810

9 798993 263809